T0202775

MONOGRAPHS ON
STATISTICS AND APPLIED PROBABILITY

General Editors

**D.R. Cox, D.V. Hinkley, N. Keiding, N. Reid,
D.B. Rubin and B.W. Silverman**

Multivariate Dependencies
Models, analysis and interpretation

D.R. Cox

Nuffield College
Oxford, UK

Nanny Wermuth

University of Mainz
Germany

CRC Press
Taylor & Francis Group
Boca Raton London New York

CRC Press is an imprint of the
Taylor & Francis Group, an **informa** business

A CHAPMAN & HALL BOOK

CRC Press
Taylor & Francis Group
6000 Broken Sound Parkway NW, Suite 300
Boca Raton, FL 33487-2742

First issued in paperback 2019

ISBN-13: 978-0-412-75410-4 (hbk)
ISBN-13: 978-0-367-40137-5 (pbk)

Library of Congress Cataloging-in-Publication Data

Cox, D.R. (David Roxbee)
    Multivariate dependencies : models, analysis, and interpretation /
D.R. Cox, Nanny Wermuth.
        p. cm.
    Includes bibliographical references and index.
    ISBN 0-412-75410-X
    1. Multivariate analysis. I. Wermuth, Nanny. II. Title

QA278.C69 1998
519.5'35—dc21   98-48949
                                                                      CIP

Library of Congress Card Number 98-48949

Visit the Taylor & Francis Web site at
http://www.taylorandfrancis.com

and the CRC Press Web site at
http://www.crcpress.com

Contents

Preface

The object of this book is to describe methods for the analysis of relations between a set of variables, with emphasis largely but not entirely on observational studies in the social sciences. This involves both the study of asymmetric dependencies of one variable or set of variables, regarded as responses, on another variable or set of variables, regarded as explanatory, and also the symmetric analysis of the joint distribution of a set of variables treated on an equal footing.

We shall emphasize also the role of intermediate variables which serve as responses to some variables and as explanatory to others. This leads us naturally to the consideration of chains of relations that are represented incisively by a fully or partially directed graph.

We assume some general knowledge of statistical methods such as regression analysis in its various forms. To make the book both accessible to workers in particular applied fields and also appealing to statisticians, the mathematical style varies appreciably between sections and chapters. We have aimed to give some theoretical justification to all the ideas discussed, but by omitting the more mathematical parts we hope that the worker who has no direct interest in statistical theory as such will find enough general explanation and motivation to make the book useful.

We illustrate the discussion with research questions from the medical and social sciences but, inevitably, because of the difficulty of fully describing applications of realistic complexity, some idealization and simplification had to be used.

The details of computer packages change so rapidly that we have not attempted to make recommendations. Most of the analyses reported here can be pieced together from software available in standard packages. For special computations MATLAB was used.

Our collaboration has been supported by a Max Planck-Forschungspreis and by the Anglo-German Academic Research Collaboration Programme. We are very grateful for this assistance.

We are also extremely grateful to the researchers connected with each of the larger studies of Chapter 6 who explained to us much of the research history and the design of data collection. They are: Hans-Ulrich Gerbershagen, head of the chronic pain clinic in Mainz, who enhanced basic research into chronic pain, designing an important scoring method for chronicity of pain; Heinrich Giesen, educational psychologist at the University of Frankfurt, who planned and organized the 15-year cohort study of several thousand pupils which involved monitoring their performance in university degree programmes; Günther Esser and Manfred Laucht, the two leading psychologists active in planning and carrying out the child development study at the Zentralinstitut für Seelische Gesundheit in Mannheim; and Carl-Walter Kohlmann, involved as the main psychologist in studies of glucose control of diabetic patients at the University of Mainz.

We thank Reinhold Streit for intensive computations and suggestions for improvement of graphical displays, Nicolai Klessinger and Nicole Geske for further computational help; and Paola Vicard for valuable comments on an earlier complete version of the manuscript.

D.R. Cox and Nanny Wermuth
Oxford and Mainz
November 1995

CHAPTER 1

Introduction

1.1 Preliminaries

Statistical considerations may enter all phases of an investigation from the formulation of objectives, the design and piloting of a study, the collection of data and the checking of data quality, to the final presentation of conclusions.

We shall be concerned with studies typically of the following kind. There are a number of individuals, people, firms, countries, etc. On each individual several variables are recorded and these vary appreciably between individuals. The objective is to describe, analyse and interpret that variation. As will emerge when we discuss examples, this formulation is sufficiently rich to cover a wide variety of investigations.

The problems of interpretation differ somewhat between the *primary analysis* of data collected for the specific research purpose under study and the *secondary analysis* of data collected for some other purpose, especially, for example, the analysis of data collected for administrative reasons. More broadly, the depth and elaboration of analysis that are sensible must depend on data quality and relevance.

Here we shall concentrate largely on methods for the primary analysis of data where the research objectives that led to the investigation are likely to play a central role.

1.2 Types of investigation

It is useful to distinguish between the following concepts:

1. In *experiments*, the system under study is wholly or largely under the investigator's control. Typically there are several possible treatments, and for each individual one treatment is chosen by the investigator, applied to that individual, and one or more measures of response are observed. The object is to compare the effect of the treatments on response.

2. In *longitudinal prospective studies*, initial measurements are made on a sample of individuals, and those individuals are followed forward in time to observe relevant outcomes at one or more future time points.

3. In *longitudinal retrospective studies*, observations are obtained on a sample of individuals at a single time point and then the history of those individuals is examined, usually with the objective of explaining the current status.

4. In *cross-sectional observational studies*, a number of variables are measured at a single time point on a sample of individuals in a system.

Illustrations. Experiments are the primary method of investigation in the laboratory sciences and in some technological fields, such as the process industries. One important type of experiment involving human subjects is the *randomized clinical trial*. Here an individual is a patient assigned at random by the investigator to one of a number of possible treatments and followed to observe a response, such as treatment success or survival within a specified period versus treatment failure or death.

In a related prospective observational study, treatment regime would be recorded for each patient but would be outside the investigator's control. Patients would be followed to observe relevant outcomes. Such studies are sometimes called *cohort studies* or *panel studies*. A further example is a voting study in which baseline data are obtained for a sample of voters and then voting behaviour is recorded over a series of future elections.

In comparable retrospective studies, disease state would be observed at one point and then variables measured in the past history of the patient or, in the case of electoral studies, questions asked about voting in previous elections as well as relevant socio-economic variables.

In a similar cross-sectional study only behaviour at one election would be involved.   □

The absence of direct control by the investigator, and more especially of randomization, means that in observational studies there is always some additional uncertainty over the genuine comparability of groups of individuals, i.e. as to whether the estimation of contrasts of substantive interest is biased.

Particular methods of analysis may be appropriate for some or all of these types of study although the difficulties of interpretation are different, with experiments usually having the least ambiguous interpretation and retrospective studies being the most prone to bias. In broad terms cross-sectional studies tend to be less informative than roughly parallel longitudinal investigations.

There are other important aspects of design, such as the design of instruments, considerations of the size of investigation appropriate and of the balance of effort between different phases of a study and of the steps desirable to achieve high data quality and completeness. Perhaps most importantly, there is the choice of individuals and treatments for study. All these matters are central to successful research, but will not be discussed in detail in this book.

1.3 Types of variable

The variables measured on each individual can be classified in a number of ways. One is by the structure of possible values, which may be effectively continuous, numerical but discrete, ordinal, nominal or binary. Qualitatively the kinds of question to be asked do not depend on this particular distinction although the precise formulation and techniques for statistical analysis are different in the various cases. We shall emphasize the two extreme cases of continuous and binary variables. The nominal case can often be dealt with by extension of methods for binary variables. The most difficult case for general treatment is probably that of ordinal variables, these being neither continuous nor nominal; we shall discuss ordinal variables in more detail in Section 3.9.

A second distinction, which has more bearing on the objectives and interpretation of the analysis, is that between response, intermediate and explanatory variables. *Intermediate variables* are treated as responses to some variables and as explanatory for other variables.

In discussions of a particular investigation, for example in the planning phase, it is sometimes sensible to begin with the explanatory variables and to ask what responses are likely to arise to them, and in other cases to begin with the responses and to ask what explanatory variables are most likely to be relevant.

Illustrations. In studies other than cross-sectional ones, the relation between variables will often be decided by the temporal order. For example in a randomized clinical trial or prospective study, measures of health status at entry would be explanatory, health status during the study an intermediate variable and overt reaction to treatment a response. If measurements of, say, diet before entry were available, then for some purposes aspects of health status at entry might become an intermediate variable.   □

In many investigations broad groupings of individuals may be used as, in effect, explanatory variables, these groupings serving as surrogates for a range of features to which it may not be possible to assign separate interpretations.

Illustration. In an epidemiological study socio-economic class might be used as a surrogate for a range of life-style and other features.   □

In a study with various objectives, variables may change status between different phases of the analysis. In other contexts where the measurement of the response of primary interest is difficult, intermediate variables may be used as surrogates for the response of ultimate interest.

Illustrations. Blood pressure after treatment may sometimes be used as a surrogate for cardiac critical events, it being assumed provisionally that improved blood pressure control will decrease the chance of a critical event. In studies of AIDS, immunological markers, such as CD4 count, may be used, again provisionally, to assess the effectiveness of treatment.   □

As already in effect noted, any distinction between types of variable in cross-sectional studies must come from substantive, i.e. subject-matter, sources. In other types of study the 'arrow of time' picks out certain relations as directed. Even in such studies, however, variables referring to the same time point must either be treated on a symmetric footing or distinguished on substantive grounds and such distinctions are typically provisional and possibly controversial. It will, however, be central to our discussion of problems with a number of variables that ordering of variables from explanatory through to ultimate response should be exploited wherever possible to aid subject-matter interpretation.

Illustration. In a study in which stress at work, alcohol consumption and general psychological well-being are among the variables measured, it might well be controversial to impose an ordering among these variables.   □

A further special matter that needs attention in analysis arises when a number of different measures are recorded addressing the same feature; these are sometimes called *indicators* for the underlying concept. This can happen both with response and with explanatory variables.

Illustration. In a study in which smoking behaviour is likely to be an important explanatory variable, but is not the main focus of the investigation, a number of different pieces of information may nevertheless be collected covering the amount of smoking, the history of smoking and the method (cigarettes, pipe, etc.; tendency to inhale). These will usually have to be condensed into a single score, by some mixture of previous experience and preliminary analysis of the data. If, however, smoking behaviour as such were of primary concern the approach would be different, more emphasis being placed on trying to isolate which aspects of smoking behaviour are of significance.   □

Particularly in relatively more complicated problems, we shall emphasize a diagrammatic representation of the relation between variables.

In this each variable is represented by a point (*node*). If the variables are to be treated on an equal footing, all are placed in a single box; see Figure 1.1a. If some are responses and others explanatory, variables on an equal footing are placed in the same box, variables of different types being in separate boxes with responses to the left of explanations; see Figure 1.1b. Next, connections between node pairs (*edges*) are drawn. *Arrows* (directed edges) point from explanatory variables to response variables and *lines* without arrowheads (undirected edges) join variables within the same box. Later we introduce two crucial further aspects. We allow edges of two types, full and dashed and, most importantly, we specify when an edge can be omitted from the graph, such an omission representing independence in a precise sense to be defined. For our initial example, with six variables, two of each of the three types, shown in Figure 1b, all possible lines are inserted, some arrows being missing.

There are further distinctions to be made, especially concerning explanatory variables. Some are *treatments*, or in observational studies are perhaps better called quasi-treatments, in the sense that at least conceptually they could be manipulated for any particular individual to be different from the value actually realized.

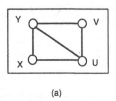
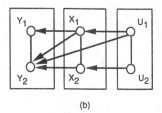

(a)   (b)

Figure 1.1 *Outline graphical representation.*
(a) Four variables, Y, X, V, U, treated symmetrically. (b) Six variables:
Y_1, Y_2 pure response variables; X_1, X_2 intermediate variables; U_1, U_2 explanatory variables; five arrows missing. Undirected lines for symmetric associations within boxes. Edges between boxes are arrows pointing to responses. Boxes indicate that directed edges present are also substantively important.

Other explanatory variables are *intrinsic* in that they are best regarded as immutable properties of the individual in question. A third type describes broad groupings of individuals into blocks, centres, countries, etc. not identified by a small number of well-defined features. We call such variables *nonspecific*. These distinctions depend strongly on the context.

Illustrations. In what for humans would be an inevitably observational study of the effect of alcohol intake during pregnancy on the live weight of the newborn, measures of alcohol intake, smoking behaviour and diet are treatment variables, the age of the mother and how often she has previously given birth (parity) would be intrinsic variables and gestation time an intermediate variable. In most contexts gender of the subject is an intrinsic variable; an exception is in studies of possible discrimination in employment, where the central issue is what the earnings of an individual would have been had the individual been male, say, rather than female, other relevant aspects of the individual remaining the same. In an educational study of children, class, school, geographical region are nonspecific variables.   □

A final distinction is between observed and *hidden* or *latent*, i.e. unobserved, variables. The latter are of two broad types. The first are the notional 'true' values of variables recorded with error. The second are hypothetical concepts which are often related to several observed variables and which are assumed to represent underlying features of the system under study.

Illustrations. Blood pressure, even if carefully measured under standardized conditions, shows substantial short-term variation for a particular person. For some purposes it may be helpful to regard recorded blood pressure as only an estimate of an underlying notional 'true' blood pressure.   □

Virtually all variables are subject to measurement error, although sometimes this may be assumed negligible compared with other sources of variation or, in the case of response variables, treated as part of the general random variation.

The constructs used in factor analysis, or in its generalizations, are examples of the second kind of hidden variable.

In our discussion we have written as though the primary observations for analysis were variables directly recorded by the investigator, or directly available where secondary analysis of data is involved. Quite often, however, the variables in terms of which interpretation is best done are *derived variables* in the sense of being combinations of more basic variables. These may be combinations determined a priori or, less frequently, may be derived by analysis of the data under study.

Illustrations. Psychological test scores are obtained from the answers to a considerable number of questions, usually by summing scores from the individual questions, possibly with differential weighting of different questions. Determination of scoring systems is itself a major statistical problem but in many applications these scores are taken as fixed either by convention or by prior investigation.

Blood pressure is conventionally measured by two components, systolic and diastolic. Often only one of these is analysed but a more sensitive approach may sometimes be to analyse two derived variables, such as the sum and difference of the logs of the two values. It would be possible to determine for each set of data that combination of the two values which provides the most sensitive analysis and this would be an example of a derived variable found from the data. Disadvantages of this approach are not only its complexity but also, more seriously, that communication between studies would be handicapped by not using common variables for interpretation.   □

In most studies of realistic complexity, especially but not only in the social sciences, there will be several primary response variables of interest. There will also frequently be several intermediate variables. In some situations it may be best to treat the different primary response variables individually as *univariate responses*. The only complication from the multivariate character of the responses then arises if a null hypothesis of, say, no treatment effect is of interest and there are different sets of explanatory variables for the various responses. Then, separate univariate analyses might be inadequate as a summary of the information in the data and a study of the distributions of *joint responses* is required, as described in detail in Chapter 5.

An allowance for selection would be needed if attention were focused on a response variable solely because it showed the most striking effect, but we try to avoid the need for such calculations.

Illustrations. In a secondary prevention study of hypertension, response variables, in that field called *end-points*, include death from heart attack, occurrence of a severe further heart attack and death from all causes. There might well be additional response variables connected with health status and adverse reaction to medication. All these variables, especially the first three, would initially best be analysed separately, with interrelationships between response variables studied at a later stage.

In a psychological study, response variables might be measures of state anger and state anxiety of subjects in various situations, taking the trait variables as explanatory. While separate analyses of the two responses would be of interest, it is entirely possible that relevant information is contained in their joint distribution, conditional for each state variable only on its own trait variables, in which case a joint analysis is required; see Section 5.3.   □

1.4 Interaction

An important role is played in many investigations of response variables by the notion of *interaction*, taken in the statistical sense that the effect on response of one explanatory variable depends on the level of one or more other explanatory variables. We comment here on the simple case of *two-factor interaction*, i.e. where a response and just two explanatory variables are involved.

The substantive interpretation is of two broad kinds depending on whether the explanatory variables are both treatments or one is a treatment and one an intrinsic variable.

Illustration. Interest may lie in the effect on response of two drugs, each either given in a standard dose or absent. There are thus four treatment combinations $(0, 0)$, $(0, 1)$, $(1, 0)$ and $(1, 1)$, where, for example, $(0, 1)$ means zero dose of drug 1 and a standard dose of drug 2. Interaction of the two treatments on the response, whether the study is observational or experimental, means that in some sense the effect of drug 2 is different at the two levels of drug 1, and vice versa. Extreme cases arise when there is no effect unless both drugs are present, i.e. $(1, 1)$ has a response different from the other three combinations, and when both drugs have an effect on their own and none when in combination. More generally, dose could be used as a quantitative variable not restricted to just two levels.   □

Illustration. With just one drug an interaction with an intrinsic explanatory variable would arise if the effects on men were different from those on women, gender being taken as the intrinsic variable. As another example, if the drug were administered in a range of dose levels and the intrinsic variable is the volume or surface area of an organ, say the heart, we would be considering possible interaction between a quantitative treatment variable and a quantitative intrinsic variable.   □

The above discussion is essentially qualitative. The usual quantitative definition of no interaction for continuous responses is that treatment differences are constant. Thus if the response were transformed, for example by taking logarithms, the definition on the transformed scale would be different. Conversely if in fact ratios of responses were constant there would be interaction on the original scale of measurement removed by taking logarithms.

Our attitude to interactions is in a sense ambivalent. On the one hand, analysis and interpretation of complex systems with many explanatory variables are hard unless the number of interactions present is quite small. Further, the absence of interaction between a treatment and an intrinsic variable is some assurance of the stability of the treatment effect under study and hence, in particular, some basis for reasoned extrapolation. On the other hand, in the kinds of psychological, sociological and other rather complex fields with which this book is primarily concerned, it is usually quite unreasonable to assume that all individuals will respond in essentially the same way and the isolation of interactions clarifying this can be a crucial result of analysis.

1.5 Purposes of analysis

It is trite but important that the purpose of an investigation has a strong influence on the methods of analysis. Sometimes there is a very specific practical decision objective such as the choice of treatment to be recommended for an individual patient. The methods described in this book are directed more at the study of dependencies and associations with the objective of 'understanding' the system under study in the sense of gaining some knowledge of the generating process, of some ability to predict, and of relating the particular data under analysis to current knowledge of the field in question.

Of course deep understanding is unlikely to be achieved by a single study, no matter how carefully planned. A key issue thus concerns the way in which subject-matter knowledge or assumptions about the problem are to be introduced and this will depend on the amount of tolerably firm quantitative or qualitative theory available and on whether there are similar or related previous data with which comparison should be made even if on a rather tentative basis.

In many areas of the natural sciences there are quantitative notions which can and normally should be the basis of analysis. In the social sciences, however, background knowledge tends to be qualitative and some of the ways that this can be introduced into an analysis of how the observations might have been generated are as follows.

Firstly, there may be an arrangement of the variables in sequence, in the way already outlined, starting with the variables regarded as the responses of main interest and continuing through intermediate variables of various types to nonspecific and intrinsic explanatory variables. Note incidentally that this ordering of variables is that most natural for a retrospective study.

Secondly, there may be qualitative constraints on aspects such as monotonicity, behaviour in limiting situations and so on.

Thirdly, we may formulate *substantive research hypotheses* asserting both that certain variables are strongly related and also that particular conditional independencies hold. A formulation in part via null hypotheses of independence is valuable in pointing towards semi-quantitative models of the system under study.

Finally, there may be substantive research hypotheses involving the strength and direction of relations between both observed and hidden variables.

When the objective is a more specific one such as *prediction* in the narrow sense or the establishment of working 'natural ranges' for certain variables a difficult general issue arises. There may be a choice between a relatively simple, very empirical method making no appreciable use of background knowledge and a more elaborate approach that uses theoretical information. This issue arises in a number of fields. It may be that while the simple method is as effective as or more effective than the more elaborate model in the short term, the more substantively based approach typically performs better if there are longer-term changes in the system.

Illustration. There is some evidence that for short-term fore-casting of simple economic variables empirical univariate models may be as effective as or more effective than methods based on elaborate econometric models.   □

Quite often information on the same issue from different sources is to be combined; the sources may be different parts of the same study or different studies. This does not raise any very special issues for analysis, the main points for consideration being as follows:

1. Are the important conclusions from the different sources consistent?

2. If not, can the discrepancies be reasonably explained, for example via suitable explanatory variables, via bias in the data or via some other identifiable lack of comparability of the studies?

3. If not, should the combination of sources be undertaken at all and, if it is undertaken, how is the final combination to be calculated and its error assessed?

These issues are discussed further in Section 4.3.

1.6 Nature of models

1.6.1 Some simple models

The results of statistical analysis in one sense provide summaries of aspects of the data judged relevant for interpretation. While this important, essentially descriptive aspect should not be discarded, there are usually good reasons, for example the need to assess uncertainty, for basing analyses on explicitly probabilistic considerations.

The basic assumptions of *probabilistic analyses* are as follows:

1. The data are observed values of random variables, i.e. of variables having probability distributions.

2. Reasonable working assumptions can be made about the nature of these distributions, usually that they are of a particular mathematical form involving, however, unknown constants, called parameters. We call this representation a model, or more fully a probability model, for the data.

3. Given the form of the model, we regard the objective of the analysis to be the summarization of evidence about either the unknown parameters in the model or, occasionally, about the values of further random variables connected with the model, and, very importantly, the interpretation of that evidence.

Illustration. One of the simplest widely used models for n individuals with one explanatory variable, X, and one response variable, Y, is to suppose that observations for the n individuals in the study are such that, conditionally on all the values of X, the values of Y are independently normally distributed with mean varying linearly with each individual's value of X and with constant variance. Note that, unless the values of X have been fixed by the investigator, this is only a partial specification in that it gives just a *conditional distribution*. A full specification of the *joint distribution* would give also the distribution of the values of X but for many purposes this may be unnecessary. That is, in the fuller specification we would give the distribution of $(X_1, Y_1), \ldots, (X_n, Y_n)$. In the conditional specification we specify only that given $\{X_i = x_i; i = 1, \ldots, n\}$, Y_i is normally distributed with mean $\alpha + \beta x_i$ and variance σ^2, independently for all i. The unknown parameters are $(\alpha, \beta, \sigma^2)$. Equivalently we could write

$$Y_i = \alpha + \beta x_i + \epsilon_i,$$

where ϵ_i denotes a *random term*, normally distributed with zero mean and variance σ^2, measuring the amount of random variation present and not explained by the *systematic component* $\alpha + \beta x_i$. The deviation ϵ_i is sometimes called the *error*, although this is often a rather misleading term in that ϵ_i frequently represents natural variation in some population rather than experimental or observational error in the sense of some kind of mistake.

In a full specification it might be supposed that the pairs (X_i, Y_i) have independently a bivariate normal distribution specified by five unknown parameters, which could be taken as the above three augmented by the mean and variance of X.

In a very careful notation random variables are represented by capital letters, and their particular observed values by the corresponding lower-case letters, so that the data in this illustration should be denoted by $(y_1, x_1), \ldots, (y_n, x_n)$. Once the difference between random variables and observed values is understood, in most contexts it becomes slightly pedantic to insist on the notational distinction and we shall not do so. That is, we shall often treat the data and random variables as equivalent.

The simple regression relation is capable of almost infinite elaboration.

An important immediate generalization, which directly or indirectly is the basis for many studies involving several explanatory variables, is the model of *multiple linear regression*. Here there are q explanatory variables x, the data for n individuals taking the form

$$(y_1, x_{11}, x_{12}, \ldots, x_{1q}), \ldots, (y_n, x_{n1}, x_{n2}, \ldots, x_{nq}).$$

The simplest model has

$$Y_i = \alpha + \beta_1 x_{i1} + \ldots + \beta_q x_{iq} + \epsilon_i,$$

where the β's are called partial regression coefficients; their interpretation will be discussed in Sections 2.7 and 4.5.

Sometimes we use a more explicit notation in which, for example, β_1 is replaced by $\beta_{Y1.2\ldots q}$. This shows first the response and explanatory variable involved and then, to the right of the full stop, the remaining explanatory variables in the equation.   □

The purely empirical interpretation of the probability distribution is often rather hypothetical and better supplemented by substantive knowledge. Nevertheless the interpretation is that we imagine a population of values produced by repeating the investigation under the same conditions; probability then refers to frequency in that population, i.e. specifies what would happen in the long run. This notion and that of an underlying parameter thus aim to capture aspects of the system under study that are free from the accidental disturbances in the particular set of data under analysis. Choice of a model and of the parameters of interest are thus key aspects in achieving fruitful formulation.

1.6.2 Types of model

Models can be broadly classified as *substantive* or *merely empirical* depending on how strongly they reflect specific subject-matter considerations. As noted in Section 1.4 in many social science applications the background for a highly specific model is not available and one looks for models that are qualitatively sensible and which allow substantive research hypotheses to be critically examined.

Illustration. In enzyme-catalysed chemical reactions simple theoretical ideas lead to a relation between rate of formation of product, Y, and concentration of substrate, X, of the form

$$Y = \beta X/(\alpha + X),$$

the Michaelis–Menten equation, and this has been found to fit reasonably well a wide variety of applications. This is a *deterministic* substantive model. Given new empirical data in which X is measured accurately and Y has appreciable error of measurement, it may be reasonable to consider the probabilistic model in which

$$Y_i = \beta X_i/(\alpha + X_i) + \epsilon_i,$$

where the properties of the random terms ϵ_i of zero mean are based on previous experience supplemented by inspection of the data and might well be those of the corresponding terms in the simple linear regression model discussed above. This is a typical example of a model the systematic parts of which come from substantive considerations and the random parts of which are largely empirically based. In this example the relative efficiency of different methods of estimation can depend quite strongly on how, if at all, the variance of error depends on X_i. More realistically both X and Y may be subject to measurement error.   □

In the social sciences such detailed quantitative specifications are rarely available and, as noted above, substantive considerations enter more commonly via hypotheses of independence and in addition by qualitative knowledge about the directions and relative strengths of dependencies.

Illustration. For example in a study of hypertensive patients in which treatment regime, blood pressure and measures of health status (quality of life) are measured, a substantive research hypothesis, corresponding to a rather naive idea of how treatments work, would be that health status is conditionally independent of treatment given blood pressure. A possible initial model is therefore to allow simple empirical representations of relations between variables which reduce to the substantive research hypothesis as a special case corresponding to special values of some of the parameters.   □

Initially, in an observational study, unlike in an experiment, all variables can reasonably be regarded as random, so that the starting point for interpretation is in principle the joint distribution of all variables.

Emphasis on certain aspects of that distribution may lead to the consideration of particular marginal and/or conditional distributions; the use of conditional distributions may arise also from more technical considerations such as ancillarity, which leads to regarding the explanatory variables as fixed at their observed values, even if in fact subject to random variation; even here, however, some examination of the distribution of the explanatory variables can be desirable, for example to check on the relation with the corresponding distribution in a target population.

1.6.3 Broad aspects of models

There are several broad aspects of a probability model to be considered. Some features will be primary in that they serve to define aspects of the research questions of direct concern. The relevant parameters are called *parameters of interest*. Primary features thus attempt to encapsulate the essence of the scientific issues involved. Other aspects will be secondary, not in the sense of being unimportant, but rather as being details necessary to complete the model specification to the point where a method of analysis can be developed. Quite often assumptions of independence of sampled individuals or random terms and of distributional shape are of this kind. The parameters associated with these secondary aspects are called *nuisance parameters*.

Illustration. In the simple regression model, often the parameter of interest is the slope β, with (α, σ^2) as nuisance parameters. This is, however, a subject-matter issue. It is certainly possible that the intercept α, or more generally the expected response at a particular value x_0, namely $\alpha + \beta x_0$, is the parameter of interest and in a study in which sources of random variability are to be explored, the variance σ^2 would become the parameter of interest, although this is probably the least common of the three main possibilities.

In theoretical statistical discussions it is possible to take the parameter of interest to be a vector, for example (α, β) or even the vector of all three parameters. In applications, however, wherever feasible it makes for more incisive final interpretation to take single parameters studied separately as the objects of interest. That is, the final interpretation will hinge on the choice of an appropriate model and, if possible, separate discussion of the meaning of key parameters.   □

So far as the primary aspects of the model are concerned it is thus desirable that each unknown parameter of interest has a distinct identifiable interpretation. Often this will be that vanishing of a parameter represents a relation of conditional independence, or the equivalence of two treatments or the agreement of the current body of data with some historical data in important respects.

It is essential that the model is chosen so that the issue at stake can be addressed.

Illustration. Suppose that it is of substantive interest to examine whether the relations among a set of response variables show some nonlinearities or whether the relation between two of the components changes with the value of a third response variable. Then an analysis assuming a multivariate normal distribution is unsuitable because in that distribution all regressions are linear and the kind of interaction mentioned cannot arise.   □

It is useful to call a model *saturated*, or, more fully, saturated with parameters, when there are the same number of unknown parameters in the model as independent pieces of data to be fitted. Thus an exact fit can be achieved to arbitrary data and the adequacy of the model cannot be tested from the data being fitted.

Illustration. Suppose that we have one or more nominal variables and that the data consist of the m cell frequencies in the multi-dimensional contingency table formed from all possible combination of values. Then the multinomial distribution in which the cell probabilities are arbitrary, thus involving $m - 1$ independent parameters, is saturated. The model can be tested only by going beyond the cell frequencies, for example by examining possible time trends in data collected over a substantial time period.

If sample means and covariances of a multivariate response variable are regarded as the basic data, a multivariate normal distribution with arbitrary mean and covariance matrix is saturated. An exact fit to the data can always be achieved and the adequacy of the model can be tested only by considering functions of the data other than means and covariances, for example by introducing explanatory variables or more complex functions of the data, for example higher-order moments. □

There are many other aspects of model formulation that will arise in the rest of the book. Some balance between simplification to capture the essence of a problem and complication to ensure realism is often required. On the whole it is probably unwise to enter elaborate discussion of aspects that are secondary in the sense of not being directly concerned with the primary issues under study, a notion sometimes called the principle of minimal modelling.

Related to this is the issue of whether to start with a relatively simple model that may need to become more complicated in the light of preliminary analysis of the data or whether to start with a rather general model that, hopefully, can be simplified as the analysis proceeds. Again, it may be sensible to analyse the data in a number of separate sections, combining the results in a final stage of analysis. We discuss some of these broad issues in the following sections.

1.7 Examination of data quality

An important part of any analysis is the checking of data quality, including the detection of missing values, logically or virtually impossible values, possibly arising from data coding errors, and in general apparently anomalous features. The methods for doing such checking are largely but not entirely informal; they include the calculation of means, standard deviations, minima and maxima variable by variable and inspection of aspects of bivariate distributions such as scatter plots, differences in means, percentages, etc. The more formal methods of testing model adequacy discussed in Section 2.10 may also reveal issues best explained via data defects, or via the inclusion of individuals who would best be regarded as outside the population under study. More broadly, whenever very surprising conclusions emerge from an analysis it is always wise to check especially carefully that they are not a consequence of some defective data or of predictions for a range of values never or rarely observed.

The more formal part of the analysis centres on the formulation, checking and possible modification of a model for the data. While it is difficult to make a general formalization of this process, it is useful to review briefly some criteria for a model to be satisfactory; these criteria in turn are the basis for model criticism and improvement.

1.8 Criteria for models

Throughout we distinguish as above between the primary aspects of the model and the secondary aspects necessary to develop a detailed method of analysis and calculation of uncertainty.

The following aspects need consideration:

1. One important role of a model is to provide a link with underlying substantive knowledge. A very specific instance of this is the use of the Michaelis–Menten equation referred to above. A central question in many social science applications concerns ways of inserting subject-matter knowledge that is more qualitative.

2. It will often be desirable to provide a direct link with previous published work in the field. It is also important to think of facilitating comparison with future related work. The latter point implies that conclusions should be presented in sufficient detail that future workers can incorporate results into their analyses. Sometimes a relatively inferior method of analysis has become established in the field. Then it will usually be necessary to discuss briefly the difference between the answers obtained by old and new methods and to explain why a new method is desirable.

3. The model should give some indication of or some pointer towards a process that might have generated the data. This is related to the issue of drawing conclusions about causal interpretation; see Sections 2.12 and 8.7. While we believe it unwise to draw causal interpretations from single studies, or even from studies of one type, especially when they are observational, there are considerable advantages in analyses that may point towards

or at least be consistent with processes that might have generated the data.

4. Parameters in the primary aspects of the model should ideally have individual specific interpretations which can be directly linked to the subject-matter questions. In particular in multiple regression and associated techniques, some of the regression coefficients have such a meaning, although considerable care is needed in such interpretations.

5. Secondary aspects of the model should give appropriate descriptions of the random variation, i.e. of what is often called 'error structure', including account of randomization procedures used in experimental design or sampling and any special features involved in the selection of individuals for study or in the measuring processes. It is quite likely that the real uncertainty in conclusions is commonly underestimated. In one form or another this is because of the neglect of a source of variability, or equivalently the presence of positive correlation between errors. Thus, at least in some contexts, careful attention to the structure of the error variability is critical, in some cases because it is of intrinsic interest and in others because of the major effect on the error of the primary comparisons. On the other hand, unnecessary complication is, of course, to be avoided.

6. The model must be rich enough to capture the central features of interest. The main immediate implication of this is that it will often be essential to include key parameters, for example representing treatment effects, regardless of the magnitude of the resulting estimate.

7. The model should be consistent with the data under analysis.

The aspect most commonly emphasized in the statistical literature is the last, partly because of the need to develop relatively formal procedures for testing adequacy of fit. Now the very use of the word 'model' recognizes that an inevitably idealized representation of the real world is involved. It is therefore essential that study of lack of fit should be focused on issues of genuine concern.

Particularly with large sets of data quite small departures from / a model may be highly significant statistically. Such departures should, from the present view-point, be recorded. They may ultimately be irrelevant or too small to be of substantive importance, but these are subject-matter judgements not to be settled by some global formula. If the departures are substantively important the action to be taken may be a relatively minor modification of the secondary aspects of the model or a rephrasing of the primary aspects or, perhaps most interestingly, a change in the whole focus of the analysis.

1.9 Model selection and development

1.9.1 General issues

The general approach to be followed in model development clearly depends greatly on the amount of quantitative and qualitative background knowledge available, including experience with similar investigations. There are two rather extreme cases. In one there is available what we shall call a *default analysis*. That is, there is a model and method of analysis, not excluded by preliminary analysis of the data, which is consistent with the general requirements listed above and which, in particular, previous experience of the field has shown to be often a reasonable basis for analysis and interpretation. In such cases we start with this analysis in the absence of specific reasons to the contrary, for example subject-matter arguments for some other model.

In other cases, typified by most of the applications in Chapter 6, no default analysis is available, although even here incorporation of background information and exploitation of relatively standard analyses as components are needed. The absence of a simple default analysis in these more complex examples stems partly from the unbalanced character typical of observational data and partly from the need to develop formulations empirically as part of the process of analysis in the absence of quantitative subject-matter theory.

We first discuss this issue in a little more detail. Then the role of relatively automatic methods of model selection and comparison is discussed.

1.9.2 Default analysis

Although it is rather remote from the main theme of this book, we consider briefly, as a familiar example where there is as a default analysis, a balanced replicated randomized factorial experiment with a quantitative response. Here *unit-treatment additivity* is the assumption that the response on a particular unit is the sum of a constant characteristic of the unit and a constant characteristic of the treatment applied to that unit. That is, there is no important interactive effect with a hidden explanatory variable.

Under unit-treatment additivity, the detailed form of randomization used implies an analysis of variance with appropriate error mean squares for different contrasts of treatment effects. Specific subject-matter considerations will indicate treatment contrasts of special interest, ideally breaking all main effects and interactions into single interpretable degrees of freedom. This is the reflection of the principle of identifying single parameters of specific interest.

An initial analysis, in this approach, would isolate mean squares of interactions of all orders and would not begin by assuming, for example, that all interactions or all three-factor and higher-order interactions are negligible. We stress that the analysis of variance is not in the present context a consequence of a somewhat arbitrarily assumed linear model but rather a deduction from the randomization used in the design plus an assumption of unit-treatment additivity common to all designs.

Thus, in this type of situation, after preliminary inspection for gross errors, one would proceed to a standard analysis. This may turn out to be the basis for a final interpretation or may have a largely exploratory role.

There are several possible reasons for going on to a different analysis.

1. Unit-treatment additivity is affected by nonlinear transformations of the response. Inspection of interactions or of residuals may suggest transforming the response, or more generally forming a derived response by combining different component response variables.

2. Presence of many appreciable interaction terms, especially ones of high order, may suggest either the need for transformation, as already indicated, or more radically abandonment of the factorial specification. Thus if for three of the factors, say A, B, C, main effects, two- and three-factor interactions are appreciable and roughly equal, summarization of the effects may best be achieved by specifying a small number of cells in the $A–B–C$ cross-classification which have responses different from the majority of cells.

3. Different specifications of error structure may be proposed, for example via a time-series or spatial-process model. Necessary conditions for this to be wise are that appreciable gain in pre-

cision results or that the error structure is of intrinsic interest. Another possibility is that departure from simple unit-treatment additivity may affect error structure. For example, in many situations error of treatment contrasts may be assessed from treatment by replicate (e.g. blocks) interaction. If partition of the treatment contrast into single comparisons isolates one especially large such comparison, a comparable decomposition of treatment × replicate interaction is conventionally examined to cover the common possibility that large comparisons have larger errors.

The essential point, however, is that for a quantitative response a full analysis of variance, i.e. one with all possible comparisons isolated, can often be used both as a basis for ultimate simplification and, importantly, if less commonly, for suggesting radical inadequacies in the initial formulation.

Similar approaches can be used in relatively simple cases of logistic regression, log-linear models of Poisson variables and other instances of generalized linear models. The method is, however, impracticable in cases of even modest complexity because non-orthogonality means that isolating separate log-likelihoods, generalizing sums of squares, for the various interaction terms cannot be done independently of other terms in the model. The issue is not so much a computational one as one of presenting conclusions in an accessible form. This use of a rather general default analysis in the way conventional in the standard work on balanced experiments is largely, although not entirely, a backward procedure. We start with a rather general model and usually hope for a final interpretation in which many of the more complex contrasts can be disregarded. Of course it would be possible to begin by assuming that, say, only main effects are present or that all interactions except for two-factor ones are negligible. This, while sometimes justifiable, sacrifices much of the scope for model criticism inherent in a full analysis of variance, for example the detection of anomalous cells. For the observational studies such as described in Chapter 6 backward procedures are typically not feasible.

1.10 Bibliographic notes

The general issues discussed in this chapter arise very widely in fields making extensive use of statistical methods. The statistical literature, however, tends to concentrate on the development of new statistical methods and on formal theories of inference rather than on broader conceptual issues. Cox (1990), Cox and Snell (1981) and Chatfield (1988) are among those discussing some of the broader issues involved in applications.

Rosenbaum (1995) has a wide-ranging discussion for observational studies; for an important early account see Cochran (1965).

Section 1.3. Stevens (1950) classified scales of measurement in a rather different way than that used here; Duncan (1984) has an interesting critique.

Section 1.4. Interactions figure prominently in the traditional literature on analysis of variance, going back to Yates (1935) and the earlier work of R.A. Fisher. They arise also in the study of high-dimensional contingency tables (Bartlett, 1935; Birch, 1963; Darroch, 1974). Various kinds of interaction in response models and their interpretation are discussed by Cox (1984).

Sections 1.6–1.9. Box (1960), Cox (1990) and Lehmann (1990) discuss the nature of statistical models. King, Keohane and Verba (1994) develop the connection with so-called qualitative methods of investigation in the social sciences. The iterative nature of model development and checking was particularly stressed by Box and Jenkins (1976) in their work on time-series analysis.

There are a large number of books describing regression models and associated methods; see, for example, Weisberg (1980), Birkes and Dodge (1993), Jørgensen (1993) and, for a detailed account of generalized linear models, McCullagh and Nelder (1989) or Fahrmeir and Tutz (1994). Box (1966) and Berkson (1946) discuss some important issues of interpretation.

CHAPTER 2

Aspects of interpretation

2.1 Preliminaries

We have already in Section 1.3 sketched the use of graphs to represent key features of the relations between variables. In the present chapter we develop this idea in more detail, concentrating on those aspects that are needed later in the book and not attempting to give a systematic account. We then consider a number of other aspects of the interpretation of models and parameters.

A central notion throughout is that of independence in the statistical sense. When we have just two random quantities, say X and Y, we say that X is independent of Y, if the distribution of Y is unchanged when we are given information about X, in particular when we are given the actual value of X. That is, X is useless for predicting Y; it can be shown that this implies that the distribution of X is unchanged if the value of Y is given, so that the apparent asymmetry in the definition is illusory. With just two variables under consideration we then say that X and Y are *marginally independent* and write $X \perp\!\!\!\perp Y$. Suppose now that there is a third variable Z and that X and Y become independent once the value of Z is given. Then we say that X and Y are *conditionally independent* given Z and we write $X \perp\!\!\!\perp Y \mid Z$. The same definition would hold if Z were a set of variables rather than just one.

An important feature of the use of graphs is the employment of two kinds of edge corresponding to two different choices of conditioning set when examining relations of conditional independence between variables. We shall also later in the book find it convenient to have two types of node, represented by *circles* for continuous variables and by solid circles (*dots*) for discrete variables, although for our initial discussion we use only the former type of node. We shall also have to introduce a number of additional conventions.

In this book, we use the terminology of graphs for a number of different purposes, namely

1. to clarify the distinction between different kinds of model;

2. especially for situations of medium complexity, to express simply both prior hypotheses of conditional independence and also hypotheses formulated after analysis of the data;

3. for exposition of conclusions;

4. for the deduction of derived properties.

By the last of these we mean that, starting from one graphical representation, we may wish to deduce another involving either the same variables regarded differently, for example by changing some roles as explanatory or response variables, or by examining a somewhat different set of variables. It is only this fourth task that involves substantial theoretical results; in the other three tasks the notation of graphs is used largely descriptively. We shall develop first the descriptive ideas, leading on to some of the theoretically more difficult issues. Thus the emphasis in the use of graphs in this book is rather different from that in some other kinds of application, for example from the important role of graphs in probabilistic expert systems.

As mentioned previously, variables are represented by nodes. Edges joining pairs of nodes may be directed, pointing from an explanatory variable to a response, or may be undirected, i.e. may treat the two variables concerned symmetrically. Absence of an edge between two nodes represents some kind of conditional or marginal independence between the two variables, the precise nature of which depends, in particular, on whether the edges in the graph are drawn with full or dashed edges. There is at most one edge between any pair of nodes.

In the definition and development of these graphs the key notion is thus in essence that of the absence of an edge as implying some form of independence. Presence of an edge might thus seem to place only the weak requirement that independence is not asserted or established. An important general point in relating theory to applications is that in applications where we represent substantive research hypotheses presence of an edge is typically a statement that a particular dependence big enough to be of subject-matter concern has occurred and this is a stronger notion not well captured in the statistical theory. In the applications to be discussed in Chapter 6 we use the presence of an edge in this much stronger sense so that it is an explicit assertion of important dependence.

This point is related to the broader matter that we do not at-

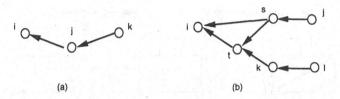

Figure 2.1 *Directly and indirectly explanatory variables.*
(a) Y_j directly explanatory for Y_i, Y_i direct response to Y_j; Y_k indirectly explanatory for Y_i, Y_i indirect response to Y_k. (b) Y_j, Y_k, Y_l all indirectly explanatory variables for response Y_i; Y_i direct response to Y_s, Y_t; path from l to j via k, t, s not direction-preserving.

tempt to attach numbers or marks to each edge to indicate the strength of relation involved.

We put considerable emphasis on arranging the variables in boxes, partly to express relations between explanatory and other types of variable and partly to express the stronger notion that the presence of an edge asserts a particular kind of dependence of substantive importance.

2.2 Some terminology

We use a terminology for the relation between nodes that is as far as possible indicative of the statistical interpretation. Thus if a node i has an incoming arrow from node j we say that Y_j is *directly explanatory* for Y_i or Y_i is a direct response to Y_j (or, in a terminology common in graph theory, j is a 'parent' of i and i is a 'child' of j). A *path* is a sequence of nodes with an edge present between each pair of nodes adjacent in the sequence. In a *direction-preserving path* each edge (i_r, i_{r+1}) is directed with an arrow pointing to i_r from i_{r+1} for all r. See Figure 2.1 for illustrations.

If there is no direct arrow from j to i but there is a direction-preserving path from j to i then we say that Y_j is *indirectly explanatory* for Y_i or Y_i is an indirect response to Y_j (or that j is an *ancestor* of i and i is a *descendant* of j). At this stage it is often convenient to label the nodes $i, j, \ldots$ rather than by the names of the random variables, such as $Y_i, Y_j, \ldots$.

A path from a node back to itself is called a *cycle*, a *directed*

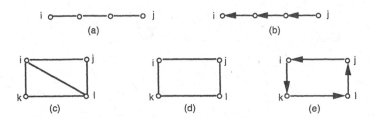

Figure 2.2 *Selected types of path.*
*(a) Undirected path between i and j. (b) Direction-preserving path to i
from j; i descendant of j; j ancestor of i. (c) Undirected 4-cycle. (d)
Chordless 4-cycle. (e) Directed 4-cycle; not permitted in chain graphs.*

cycle being a cycle that is direction-preserving. A *chordless cycle*
is one in which the only edges present are between adjacent nodes
forming the cycle. See Figure 2.2 for a key to the definitions of
paths and cycles.

A graph is called *complete* if all edges are present. A subset
of nodes together with the associated edges, i.e. ignoring edges
involving nodes outside the subset, is called an *induced subgraph.*
For some purposes an important role is played by the *cliques* of the
graph, a clique being defined as a maximal subset of nodes such
that the subgraph induced by them is complete. That is, a clique
is a complete subgraph which would turn into an incomplete graph
if even one more node were added; see Figure 2.3. Cliques play a
key role in building up joint distributions as products of factors,
although we shall not make explicit use of that here.

In an undirected graph the sets of nodes A and B are *separated*
by C if every path between a node in A and a node in B passes
through at least one node of C. An extension to partially or fully
directed graphs is given below.

It is helpful to distinguish a number of situations involving a
pair of nodes i and j in a directed graph, each directly connected
to another node t, i.e. with t being a *common neighbour:*

(i) there may be a common response Y_t to two explanatory variables Y_i and Y_j so that t is a *common sink* node in the path
$i \longrightarrow t \longleftarrow j$;

(ii) there may be a common explanatory variable Y_t to two variables Y_i and Y_j so that t is a *common source* node in the path
$i \longleftarrow t \longrightarrow j$;

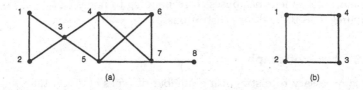

Figure 2.3 *Two illustrative graphs, each having four cliques.*
(a) Cliques are $(1,2,3); (3,4,5); (4,5,6,7); (7,8)$. *(b) Cliques are*
$(1,2); (2,3); (3,4); (4,1)$.

Figure 2.4 *Permissible types of paths via single or connected joint re-*
sponses.
Sink-oriented (a) V-configuration; (b) U-configuration of full edges; (c)
U-configuration of dashed edges; (d) U-configuration of dashed arrows
and full lines. Path end-points may be connected, but no other types of
path via response nodes in chain graphs are considered in this book.

(iii) variable Y_j may be indirectly explanatory for Y_i via an intermediate variable Y_t so that t is a *transition node* in the path
$i \leftarrow t \leftarrow j$.

In these definitions there may or may not be an edge directly between i and j. An important possibility, however, is that there is no
such connecting edge. Then we speak of a *directed V-configuration*
when one of the above three possibilities is present, or of a *sink-*,
source-, and *transition-oriented V-configuration*, respectively.

A common neighbour at which two arrows meet head on, as
in Figure 2.4a, is also called a *collision* node. For two unjoined
nodes with a common collision node, that is for a sink-oriented
V-configuration, the onterpretation will be that there are two explanatory variables, marginally or conditionally independent, having a common response. In a *sink-oriented U-configuration* the arrows point from unjoined nodes i, j to nodes t_1 and t_2 joined by a
path of undirected edges. That is, there are two marginally or conditionally independent explanatory variables Y_i and Y_j with two

or more dependent responses corresponding to nodes $t_1, t_2, \ldots$; see Figures 2.4b to 2.4d for illustrations.

2.3 Types of graph

It is necessary to distinguish a number of types of graph that will arise in the following discussion. We first give a very broad classification of graphs and then list some rather more specific types that arise in applications. Of these by far the most important in the context of this book are the joint-response chain graphs and their special case, the univariate recursive regression graphs.

A graph without boxes specifies the part of the statistical model giving the set of independencies holding. A graph with boxes incorporates assumptions or information about the relation between variables as purely explanatory or purely response or intermediate, with the added requirement that each edge present in the graph corresponds to a specific pairwise (conditional) association of interest.

We now list a number of different special kinds of graph. The first two give simple descriptions for a set of variables treated on an equal footing.

Concentration graph. Here edges are undirected full lines. Absence of an edge between nodes i and j means that the variables Y_i and Y_j are conditionally independent given all other variables. In the multivariate normal distribution absence of an edge means that there is a zero in the corresponding position in the concentration matrix, i.e. inverse covariance matrix, as specified in detail in Section 3.4. In a contingency table absence of an edge corresponds to the vanishing of parameters specifying two-factor and higher-order interactions in a log-linear representation of the cell probabilities.

Covariance graph. Edges are undirected dashed lines. Absence of an edge between nodes i and j means that the variables Y_i and Y_j are marginally independent. In the multivariate normal distribution absence of an edge means that there is a zero in the corresponding position in the covariance matrix and for a contingency table absence of an edge means lack of two-factor interaction in a marginal two-way table.

In the applications we shall consider, insertion of the distinction between explanatory and response variables plays a central role. We give next the simplest and, where achievable, the most satisfactory representation of this.

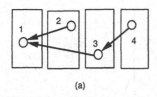
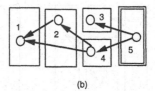

(a)   (b)

Figure 2.5 *Graphs with single responses and single intermediate variables within each box.*
(a) Simple univariate recursive regression graph; in regression of Y_1 on Y_2, Y_3, Y_4 the coefficient of Y_4 vanishes so that Y_1 is independent of Y_4 given Y_2, Y_3; also Y_2 is independent of Y_3, Y_4. (b) Generalized univariate regression graph; stacked boxes for nodes 3 and 4 denote Y_3 independent of Y_4 given Y_5; double lines around node 5 indicate Y_5 fixed at its observed values, i.e. the distribution of Y_5 not specified by the model.

Univariate recursive regression graph. Here the nodes can conveniently be arranged in a line of boxes, with one node within each box. The key requirement is that the variable in any box is a response to variables in boxes to its right. All edges are directed full arrows pointing from right to left. Absence of an edge from j to i, $i < j$, means that Y_i is independent of Y_j given Y_k, $i < k \neq j$. In the multivariate normal setting this means that in the linear regression of Y_i on all variables to its right, the coefficient of Y_j vanishes; see Figure 2.5a. This type of system is especially important both for its directness of interpretation and for the simplicity of its analysis via a sequence of univariate analyses.

Generalized univariate recursive regression graph. We consider a relatively minor generalization. While there may be two or more variables to be treated on an equal footing, these variables are assumed conditionally independent given all variables to the right. The variables in question are then shown in vertically stacked boxes; see Figure 2.5b. The same simplifications of analysis and interpretation continue to apply. We sometimes use the additional convention that the box to the far right, if it corresponds to a single variable or to several variables that are purely explanatory, is enclosed in double lines. Their values are then regarded as fixed by design or for analysis, i.e. no attempt is to be made to represent their probability distribution. Nevertheless, we report their relations as observed sometimes by undirected edges.

Next we define a *skeleton* graph as a graph without boxes ob-

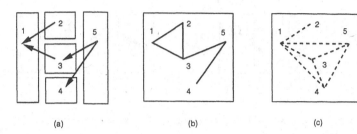

(a)   (b)   (c)

Figure 2.6 *Illustration of a directed graph generating undirected graphs with more edges for joint distribution.*
(a) Generating graph; fully directed in five nodes. (b) Corresponding induced concentration graph; an edge present means pairwise association given all other variables; for instance, the additional edge (2, 3) obtained due to conditioning on common sink node 1. (c) Corresponding induced covariance graph; an edge present means pairwise marginal association; for instance, the additional edge (3, 4) obtained due to marginalizing over common source node 5.

tained by discarding information about the type of edge, i.e. about whether it is full- or dashed-line, directed or undirected. In effect the skeleton graph specifies the variables under study and the pairs of variables expected to be related regardless of the specific type of relation. An interesting question concerns the conditions under which the interpretation of a graph changes by adding or removing the orientation of some of the edges.

Figure 2.6 and the fuller discussion in Sections 2.4.4 and 8.5 illustrate the relations between a given directed graph and the corresponding dashed- and full-line graphs induced for the joint distribution.

Directed acyclic graph. Now suppose that while the variables are not arranged in boxes, all edges are full-line and directed and that there are no cycles, i.e. it is not possible starting from one node following the direction of the arrows to return to the original node. This means that no variable is to be taken as explanatory to itself; see Figure 2.7a. One interpretation of the graph can be obtained by deleting for a given node i all responses to it. Then variable i is conditionally independent of all other variables given the variables directly explanatory for i.

A directed acyclic graph can be converted, in general though not uniquely, into a univariate recursive regression graph. For this,

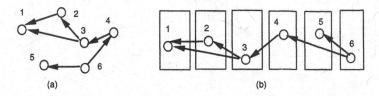

(a)   (b)

Figure 2.7 *Directed acyclic graph and compatible regression graph.*
(a) Directed acyclic graph; for instance $Y_2 \perp\!\!\!\perp (Y_4, Y_5, Y_6)|Y_3$. (b) One
corresponding univariate recursive regression graph; node 5 could be any-
where to the left of node 6 without changing the independence structure.

one way is to number first a node which occurs only as response,
to ignore it temporarily and then to work backwards in the same
way through the remaining subgraph; see Figure 2.7b for a graph
derived in this way from Figure 2.7a. Whenever there are several
variables that occur only as responses no order between these variables is implied and they could appear as stacked variables in the
graph with boxes, i.e. in the univariate recursive regression graph.
Thus boxes with nodes 4 and 5 could have been stacked in Figure
2.7b.

It is central to our discussion that the same set of independencies
can be represented in a number of different ways and that these
have different interpretations. This is illustrated in very simple
form in Figure 2.8. In Figure 2.8a, U is an explanatory variable, X
an intermediate response variable and Y a response. When X is regressed on U there is no relation, i.e. $X \perp\!\!\!\perp U$. When Y is regressed
on X, U there is a contribution of substantive interest from both
variables. In Figure 2.8b, the roles of X and U are interchanged. In
Figure 2.8c, X and U are explanatory variables on an equal footing and known for substantive reasons to be independent and we
study the dependence of Y on X and U symmetrically. Figure 2.8d
is the *underlying directed acyclic graph* for each of Figures 2.8a to
2.8c. Formal tests of goodness of fit using any of the above against
a common alternative would yield identical answers.

While for some technical purposes it may be convenient to replace the systems of Figures 2.8a to 2.8c by that of Figure 2.8d,
for an interpretation with the roles of variables as response, intermediate and explanatory given from the substantive context, the
distinctions must be preserved.

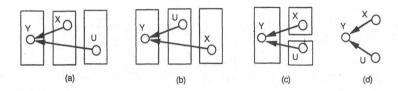

Figure 2.8 *Three regression graphs representing the same independence,*
$X \perp\!\!\!\perp U$, *and having the same underlying directed acyclic graph.*
(a) Variable Y is response, X intermediate, U explanatory; (b) roles of
X and U interchanged; (c) X and U both explanatory. (d) Underlying
directed acyclic graph.

Joint-response chain graph. Here the arrangement is again in
boxes from left to right as in the univariate recursive regression
system, but there are in general several nodes within each box.
Within each box there is either a full-line concentration graph or a
dashed-line covariance graph, each considered conditionally given
variables in boxes to the right. For a joint normal distribution
dashed lines in a given box mean that the covariance matrix of
residuals is considered and full lines mean that the concentration
matrix of residuals is taken.

Arrows pointing to any one box are either all dashed arrows or
all full arrows. Dashed arrows to a node i indicate that regressions
of Y_i on variables in boxes to the right of i are being considered,
whereas full-line arrows mean that the regression is taken both on
variables in boxes to the right of i and on the variables in the same
box as i. That is, the nature of the arrows into a box indicates the
particular regression relation under consideration; see Figure 2.9
for a simple illustration of the distinctions.

Partially directed acyclic graph. This next class of graphs in a
sense underlies joint-response chain graphs. Variables joined by an
undirected edge are treated symmetrically, so that obvious consistency conditions have to be satisfied. For example, if i is a response
to a node k and j is explanatory to k any edge between i and j
must be directed. If, for instance, all edges, directed or undirected,
are full-line, then absence of an edge between i and j implies the
conditional independence of the associated variables given all variables k directly or indirectly explanatory to either. The acyclic
property here is that it is impossible starting from an arrowhead
at one node to find a path returning to that node without meeting

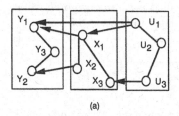 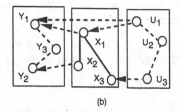

(a)   (b)

Figure 2.9 *Two joint-response chain graphs, same arrows and lines but different type of edge implying different independence statements.*
For instance, absence of an edge for (Y_1, Y_2) means in (a) $Y_1 \perp\!\!\!\perp Y_2 \mid (Y_3, X_1, \ldots, U_3)$, in (b) $Y_1 \perp\!\!\!\perp Y_2 \mid (X_1, \ldots, U_3)$; absence of an edge for (X_1, U_3) means in (a) $X_1 \perp\!\!\!\perp U_3 \mid (X_2, X_3, U_1, U_2)$, in (b) $X_1 \perp\!\!\!\perp U_3 \mid (U_1, U_2)$; absence of an edge for (X_2, X_3) means $X_2 \perp\!\!\!\perp X_3 \mid (X_1, U_1, U_2, U_3)$ in both.

an opposing arrowhead along the path. If we start with a joint-response chain graph, the associated partially directed graph is obtained by deleting the boxes.

Again there is typically more than one way in which a partially directed acyclic graph can be converted into a graph with boxes without changing the set of independencies implied. One way is to number first nodes which occur only as responses not connected by edges, to ignore them temporarily, then to number the nodes which occur only as responses but which are joined to each other by an undirected path, and then to work backwards in the same way through the remaining subgraph.

The importance of joint-response chain graphs and their role in clarifying the distinction between various models can, especially in the case of multivariate normal variables, be expressed via constraints setting to zero either elements of the corresponding concentration or covariance matrices or setting particular regression coefficients to zero.

The distinctions between what may seem at first sight a bewildering range of possibilities are needed to capture the differences between various kinds of possible research hypotheses and associated statistical models. They raise a number of issues about the relation between the different forms, and some of these we address in the next section and in Section 8.5.

In the course of that discussion we shall need as an intermediate step a further type of graph called a *summary graph* whose definition and properties we discuss later in Section 8.5.

2.4 Some more specialized properties

2.4.1 General remarks

We now outline without proof independence properties and some of the relations between the various kinds of graph introduced above. Undirected full-line graphs and directed acyclic graphs have an extensive literature, partly because of their connections with problems in statistical physics and studies of probabilistic expert systems, respectively. Undirected graphs are mostly of indirect interest in the present context, however, because in the situations considered in this book it is rare to have just a set of variables considered on an equal footing.

2.4.2 Some Markov properties

We first consider the simplification of the independencies in a full-line undirected graph for p variables, i.e. a concentration graph. Then the absence of an edge between nodes i and j means that the corresponding random variables are conditionally independent given all other $p-2$ random variables.

More generally, sets of nodes A and B represent sets of random variables that are conditionally independent given all the remaining random variables if there are no edges joining a node in A with a node in B. This conclusion can be sharpened as follows. Suppose that C *separates* A and B, i.e. that every path from a node in A to a node in B must pass through C in the sense of having a node in C. Then the random variables in A are conditionally independent of those in B given those in C. This is the *Markov property* of concentration graphs. The connection between this definition and the possibly more familiar definition of the Markov property for a stochastic process in time is that the latter can be formulated as requiring the conditional independence of the past and future of the process given the present value, that is they concern a special directed acyclic graph: a path of arrows, called a *Markov chain graph*. As a more general example, Figure 2.10 illustrates the conditional independencies associated with two different concentration graphs.

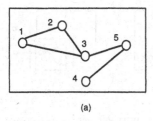 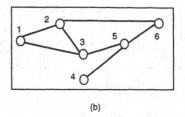

(a)   (b)

Figure 2.10 *Two concentration graphs illustrating the Markov separation property of full-line graphs.*
For instance (a) node 3 separates (1, 2) and (4, 5), nodes (2, 3, 5) separate 1 and 4. (b) Nodes (3, 6) separate (4, 5) and (1, 2), node 5 separates 4 and (1, 2, 3, 6).

The above result enables one to read off a given undirected concentration graph all independency statements implied by it. To do this requires that the probability density of the random variables concerned is strictly positive at all points; rather more generally, it is enough if the distributions are such that

$$A \perp\!\!\!\perp B \,|\, C \text{ and } A \perp\!\!\!\perp C \,|\, B$$

implies

$$A \perp\!\!\!\perp (B \cup C).$$

There is a corresponding result for covariance graphs. That is, if C separates A and B, then the random variables in A and B are conditionally independent given those not in $A \cup B \cup C$. This result, however, has been proved only under the much stronger condition that

$$A \perp\!\!\!\perp B \text{ and } A \perp\!\!\!\perp C$$

implies

$$A \perp\!\!\!\perp (B \cup C).$$

This is satisfied only for very special families of distributions, in particular the multivariate normal distributions to be discussed in the next chapter.

2.4.3 Relation between concentration and covariance graphs

If in a concentration graph any two sets of nodes A and B have no path between them it follows from the above Markov property that the variables are marginally independent, i.e. that the sets

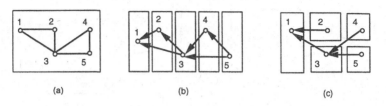

Figure 2.11 *Concentration graph and two compatible generating graphs.* *(a) Concentration graph. (b) Generating graph of the same skeleton; no sink-oriented V-configuration. (c) Generating graph with two sink-oriented V-configurations in nodes* $(2, 1, 3)$ *and* $(4, 3, 5)$; *for instance, conditioning on collision node* 1 *of the unjoined pair* $(2, 3)$ *generates an additional edge for this pair in (a).*

A and B would be unconnected in a covariance graph, because the separating set C is empty. The covariance graph will have the same skeleton as the concentration graph if the latter consists exclusively of disjoint complete subgraphs. In general, however, as soon as there is an incomplete subgraph no direct passage between properties in the two graphs can be made.

2.4.4 Derivation of concentration and covariance graphs

First suppose that we start either with a given univariate recursive regression or, for the present purpose equivalently, with the underlying directed acyclic graph. We may then ask for its concentration graph, i.e. what the set of pairwise conditional independencies given all remaining variables would be, that is if all variables were treated on an equal footing.

The new graph is constructed as follows. Whenever there is a sink-oriented V-configuration, that is there are marginally or conditionally independent explanatory variables Y_i, Y_j having common response Y_t, we add an edge between i and j. Next we produce the skeleton graph, defined previously, by deleting all boxes and ignoring the type of edge, in particular replacing directed edges by undirected ones. See Figure 2.11 for two different fully directed graphs generating the same concentration graph.

A similar procedure can be applied if we consider a selected box and all boxes to the right of it. This gives the concentration graph of the variables selected and variables possibly explanatory to them marginalizing over those to the left, i.e. over responses to either.

A corresponding question can be posed for the covariance graph. For an illustration see Figure 2.6 above. That is, we can ask which pairs of variables are marginally independent and thus not joined by a dashed edge in the covariance graph. An edge will be absent in the covariance graph, i.e. the variables will be marginally independent, if every path between them in the directed acyclic graph has a collision node. An edge (i, j) will be present if node i is a descendant of j, or vice versa, or if nodes i and j have a common source k, i.e. each of i and j is a descendant of some node k. This means that one of the variables is directly or indirectly explanatory to the other or that the two variables have a common explanatory variable, again directly or indirectly. Conditions under which variables Y_i and Y_j will then be marginally dependent are discussed in Section 8.5.

2.4.5 Markov property of joint-response chain graphs

In Section 2.4.2 we stated an important Markov property of undirected full-line graphs, i.e. of concentration graphs, that enables conditional independency statements to be read off via the condition that all paths between A and B must pass through the conditioning or separating set C. It is important to consider the corresponding condition for directed acyclic graphs and more generally for joint-response chain graphs.

The condition for C to be a separating set is slightly more complicated.

Variables of two sets A and B are independent given the variables in a set C if there are no paths between A and B of a particular kind that we term 'active', because they will induce a correlation among the variables at the path end-points under general conditions discussed in Section 8.5. A path in a directed acyclic graph is *active relative to C*:

 (i) if it is a collisionless path wholly outside C or

(ii) if a collisionless path wholly outside C is generated by it after conditioning on C, that is if along the path every noncollision node is outside C and every sink-oriented V-configuration becomes completed by a full line because the collision node in it is either in C or has a descendant in C.

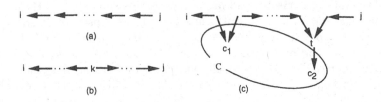

Figure 2.12 *Types of active, i.e. of correlation-inducing, paths for pair (i, j) relative to C.*
(a) Collisionless path wholly outside C, i a descendant of j. (b) Collisionless path wholly outside C; i and j have common source node k. (c) Collision path with every collision node in C or having a descendant in C, all other nodes outside of C; a collisionless path outside C is generated by conditioning on C, that is by inserting a full line for every pair with such a collision node.

In the absence of an active path A and B are separated by C and the corresponding variables are conditionally independent given C. Figure 2.12 exemplifies these possibilities, showing in Figures 2.12a and 2.12b the two relevant types of collisionless path and, in Figure 2.12c, an active collision path.

Alternatively, the separation criterion for directed acyclic graphs can be expressed in terms of a constructed full-line undirected graph to which the simple separation criterion for these graphs is applied. This criterion can be directly formulated for the more general full-edge joint-response chain graphs as follows.

First, the given chain graph is reduced by deleting nodes of descendants of A, B, C outside the union of the three sets and also all their edges. Second, the resulting graph is modified by adding a full line to complete each sink-oriented V-configuration and to connect the end-points of each sink-oriented U-configuration. Third, each arrow is replaced by a full line. The resulting graph is the concentration graph generated by the given chain graph for the subset of variables obtained by deleting descendants of A, B, C outside the union of the three sets.

Then A and B are separated by C in the full-edge chain graph if C separates A and B in the concentration graph constructed as described.

No similar results are currently available for dashed-line or mixed full- and dashed-line joint-response chain graphs. In Section 8.5 we extend the results so as to read from a directed graph whether it implies conditional dependence. In some of the applications we consider this is as or more important than obtaining independencies.

2.4.6 Decomposability of independence hypotheses

We call an independence hypothesis *decomposable* if it can in its entirety be expressed as the vanishing of regression coefficients in a univariate recursive regression process. For instance for the special family of distributions called trivariate normal for X, Y, Z the hypothesis

$$Y \perp\!\!\!\perp X \mid Z \quad \text{and} \quad X \perp\!\!\!\perp Z \mid Y$$

is equivalent to

$$(Y \perp\!\!\!\perp X \mid Z \quad \text{and} \quad X \perp\!\!\!\perp Z) \text{ or to } X \perp\!\!\!\perp (Y, Z)$$

and both are of the univariate recursive form. On the other hand, no reordering of variables will express the hypothesis

$$Y \perp\!\!\!\perp X \quad \text{and} \quad Z \perp\!\!\!\perp U$$

in univariate recursive form and hence the hypothesis is nondecomposable.

Suppose that we have obtained a representation of the independencies of a nondegenerate joint distribution by an undirected or partially directed graph. An important question concerns the representation of this as a *decomposable independence structure*, that is by an univariate recursive regression graph specifying the same set of independencies so that it is an *independence-equivalent* graph.

A full-line undirected graph, i.e. a concentration graph, can be oriented to be independence-equivalent to a directed acyclic graph if and only if it contains no chordless n-cycle, for $n \geq 4$. The reason is that every V- configuration of a full-line undirected graph can become source- or transition-oriented, but not sink-oriented, without changing the independence interpretation of the missing link. In an undirected graph with chordless n-cycle this is impossible without generating directed cycles, it is called a *nondecomposable graph*.

There is a corresponding condition for an undirected dashed-line graph, i.e. a covariance graph, provided in the joint distribution it is possible to derive appropriate conditional independencies from marginal independence, such as described in Section 2.4.2. That is, the distribution has to mimic this property of the normal distribution. The condition is that there is no chordless 4-chain. The reason here is that every V-configuration in a dashed-line undirected graph can become sink-oriented only if the independence interpretation of each missing edge is to be preserved. With a chordless 4-chain this is impossible. The condition implies that chordless n-cycles are also not permitted.

Decomposability of an independence structure does not translate directly into decomposability of a corresponding statistical model without additional assumptions about the distributions. This involves what has been called closedness of a family of distributions under conditioning or marginalizing, both satisfied for the joint normal distribution. For approximate decomposability of a statistical model the conditional and marginal distributions involved should be of similar form although possibly specified by different parameters.

2.4.7 Equivalence of independence structures in joint-response chain graphs

Another important question is whether independence structures represented by two different types of graph are identical so that formal tests for goodness of fit against a common alternative will yield identical answers.

Given two joint-response chain graphs with the same skeleton, i.e. with the same set of nodes and edges once boxes are deleted and the type of edge is ignored, we may want to know whether the set of conditional independencies involved is the same. The condition for this is currently known only for full-line joint-response chain graphs. In this case the set of (generalized) collision nodes coincide, that is, each sink-oriented V-configuration in one graph is also a sink-oriented V-configuration in the other and each sink-oriented U-configuration in one graph is also a sink-oriented U-configuration in the other.

A special case of this result is that a full-line joint-response chain graph is independence-equivalent to an undirected full-edge graph of the same skeleton if and only if the former contains neither a sink-oriented V- nor a sink-oriented U-configuration. In that case we can order the nodes in the partially directed graph without changing the independencies implied by the graph, that is in such a way that every V-configuration obtains a source- or transition-oriented form.

2.5 Hidden variables

Throughout the discussion above it is assumed that the variables represented in the graph are directly observed. Sometimes, however, it is important to incorporate hidden, i.e. unobserved, variables. As already discussed briefly in Section 1.3, this may be either because such variables are thought necessary for a good subject-matter interpretation or because puzzling features found empirically can possibly be explained via hidden variables. See Section 8.4 for more detailed discussion.

In particular the relation between undirected or partially directed graphs and directed acyclic graphs can be developed further by the introduction of hidden variables. This leads to synthetic graphs in which some of the nodes are denoted specially to show that they are not observed and that marginalization or conditioning on the corresponding variable is required to relate to the data under analysis.

To develop these ideas we replace an undirected dashed edge $i\text{---}j$ by a source orientation of an added unobserved node, i.e. by the configuration

$$i \longleftarrow \!\!\!\not\!\!\phi\!\!\longrightarrow j,$$

where $\not\!\!\phi$ denotes a synthetic node over which marginalization takes place. This essentially amounts to writing the random variables (Y_i, Y_j) corresponding to the nodes (i, j) in the form

$$Y_i = g_i(\xi_i, \eta), Y_j = g_j(\xi_j, \eta),$$

where $g_i(.,.), g_j(.,.)$ are arbitrary functions and (ξ_i, ξ_j, η) are independent random variables.

Further, we replace an undirected full-line edge $i \longrightarrow j$ by a sink-oriented V-configuration of an added unobserved node, i.e. by the configuration

$$i \longrightarrow \boxtimes \longleftarrow j,$$

where $\boxtimes$ denotes a synthetic node over which conditioning occurs.

It can be shown that if we start with any dashed-line joint-response chain graph we can obtain a synthetic directed acyclic graph, i.e. one in which specified nodes represent variables over which marginalization occurs. By using the decomposability of such a graph into univariate dependencies, such representations may be useful in suggesting ways in which the data might have been generated.

The position when full-line undirected edges are replaced is a little more complicated. A directed acyclic graph will result, implying the same independencies for the observed nodes unless in the original joint-response chain graph there is a subgraph with a sink-oriented U-configuration or there is a sink-oriented V-configuration from which a direction-preserving path of arrows leads to nodes joined by an undirected line.

The representation of dependence via marginalizing over a hidden variable is often entirely appealing. From an interpretational point of view the notion of conditioning on an unobserved feature may seem contrived. It can, however, sometimes be interpreted as representing the selection effects involved in studying a particular accessible population rather than the ideal target population. Such conditioning may also correspond in other contexts to forming a target population of interest, for example those having a specific disease regarded as a subpopulation of some larger group.

2.6 A general formulation

Graphs focus attention on conditional independencies between the variables concerned. Sometimes, however, we wish to consider the joint distribution not of the full set of random variables but of some derived set. A rather general formulation is as follows.

We start with a univariate recursive regression graph for a full set of variables Y_V, say. We partition these variables into three disjoint sets (Y_S, Y_M, Y_C) and then consider the conditional distribution of Y_S given Y_C. That is, we marginalize over the variables Y_M. It is desirable to have rules for determining the structure of this new distribution from that of the original set of variables.

An example is where Y_S, Y_M, Y_C represent respectively final response, intermediate and purely explanatory variables. For some purposes, in particular the study of the direct dependence of response on baseline variables, it is necessary to marginalize over the intermediate variables. This can be carried out directly by omitting the intermediate variables from the analysis but it can still be very helpful to read off the graph the consequences of the structure discovered for the full data.

This and similar ideas are developed further in Section 8.5.

2.7 Interpretation of regression coefficients

We now turn to some more general issues of interpretation which are related to the topic of Section 2.6 in that they are concerned with the interpretation of regression coefficients.

A primary role in many statistical analyses is played by regression coefficients expressing the dependence of a response variable on an explanatory variable. The simplest forms of regression coefficients are those in linear relations, but there are many generalizations. For example, there may be nonlinear relations such as the Michaelis–Menten equation or a relation for exponential growth up to an asymptote or models containing squared terms representing curvature. Also we encounter other forms of response variable, such as binary ones for which special forms of relation such as linear logistic regressions are often appropriate.

In the latter case regression coefficients refer to a transformed scale of response and special methods may be needed to transform the conclusions into a form more directly interpretable, for example in terms of probabilities rather than transforms of probabilities. For the rest of this section, however, we concentrate on linear regressions

$$Y_j = \alpha + \beta_1 x_{j1} + \ldots + \beta_q x_{jq} + \epsilon_j,$$

in which the response variable Y is a measurement on a well-defined scale. Such regressions are, however, of some interest in any situation in which the dependence has a strong linear component, i.e. even in some situations where detectable departures from linearity occur.

As noted in Section 1.5, we sometimes use a more explicit notation for the parameters in which the two variables directly concerned are shown as well as the other explanatory variables in the equation. In this notation β_1, say, becomes $\beta_{Y x_1 . x_2 \ldots x_q}$, or more concisely, where the context is clear, $\beta_{Y 1.2 \ldots q}$. It can be crucial for interpretation that the regression coefficient of, say, Y on X_1 can change drastically if the set of further explanatory variables changes. For instance if x_1 and x_2 are highly related, it is entirely possible for β_{Y1} and $\beta_{Y1.2}$ to have opposite signs.

Suppose initially that the explanatory variables $x_1, \ldots, x_q$ represent distinct explanatory variables. Then the interpretation of, say, β_1, the regression coefficient on x_1, is that it gives the change in $E(Y_j)$ for a hypothetical unit increase in x_1. In this hypothetical increase $x_2, \ldots, x_q$ are held fixed. A weaker but closely related interpretation is that we consider two subpopulations of individuals differing by one unit in x_1 and with unchanged values for the other explanatory variables. Then β_1 is the difference of the means of the two subpopulations. Furthermore, any other variables explanatory for Y but not included in the equation, for example because they have not been measured, change with x_1 in the same way as in the data under analysis. These conditions on the meaning of a regression coefficient are a very important constraint on practical interpretation.

The interpretation raises a number of issues. If x_1 represents a treatment or potential treatment, then β_1 is often a reasonable base for substantive interpretation. It appears to represent the average effect on response of an intervention in the system. The two qualifications noted above are, however, crucial.

Illustration. Suppose that Y is log survival time to a cardiac event and that x_1, x_2 are respectively diastolic and systolic blood pressure. Then the notion of a change, say, of 10 mm in x_1 with x_2 held fixed, is likely to be highly artificial. To assess the effect of blood pressure changes it will normally be necessary to consider some composite change in the two components or to remove one component from the model and to concentrate on the effect of the one remaining.   □

The second requirement involves omitted, including possibly unmeasured, explanatory variables. When x_1 corresponds to a randomized treatment, the omitted variables are induced to have the same distribution at all levels of x_1 and so the condition is automatically satisfied. In an observational study, on the other hand, the possibility that an apparent effect of a treatment variable arises from effects induced by unmeasured variables is one major cause for caution in interpretation. Further, if an intervention in the system changes the value of x_1 there will often be no particular reason for expecting the unmeasured variables to change in the way specified above. If β_1 is to be interpreted as in some sense the overall effect of changing x_1 on its own, then the effects consequent on induced changes in the unmeasured variables should be small in order for a simple direct interpretation to be valid.

An explanatory variable whose inclusion or exclusion has an important effect on the relation between, say, Y and x_1 is called a *moderating variable* for that relation. This includes as special cases a variable that taken with x_1 has an interactive effect on Y, and also a *confounding variable*, that is a variable strongly related to x_1 given the remaining explanatory variables.

As we have noted, the sign of the relation may well be changed even in systems in which all relations are linear, when there is a confounding variable. In extreme cases it would be possible for a series of studies all to show effects of the wrong sign because of the omission of an important confounding variable. This issue is commonly discussed under the name of the Yule–Simpson paradox and is related to problems with very unbalanced analyses of variance and to some aspects of multicollinearity. In observational studies, moderating variables that ideally should have been included but were not observed are a particular source of concern.

The interpretation of the coefficients of explanatory variables that are constructed from directly observed variables may be less direct. Thus if x_1 is a directly measured variable and $x_2 = x_1^2$, the interpretation of β_1, β_2 is initially as a pair defining a curve, probably best understood by plotting it over the relevant range of values of x_1. To interpret the parameters separately we may regard β_2 as indicating the amount and direction of curvature, the specific interpretation of β_1 depending on the choice of a reference level for x_1. More explicitly, if we fit the relation in the form

$$E(Y) = \beta_0 + \beta_1(x_1 - a) + \frac{1}{2}\beta_2(x_1 - a)^2,$$

for some suitable reference level a, possibly but not necessarily the mean of x_1 for the data, then β_1 is the slope at $x_1 = a$ or more generally the average slope over some distribution of values of x_1 having mean a, and β_2 is the second derivative of the curve.

The use of quadratic regressions in the present book is, however, largely a device for checking linearity. If on fitting the second-degree relation the estimate of β_2 is clearly significant, some modification of the linear relation is needed and this may take a number of forms; see Chapter 6 for some examples.

Rather similar comments apply when a variable is constructed to examine an interaction, say between two variables x_1, x_2, via the regression coefficient on the derived variable $x_3 = x_1 x_2$, where x_1, x_2 should be measured from suitably central origins. If both variables represent treatments it is probably best to interpret the parameters $\beta_1, \beta_2, \beta_3$ collectively via the estimated responses at four suitable base points. If, say, x_2 represents an intrinsic property of the individuals under study, such as gender or previous medical history, and x_1 a treatment variable, then the treatment effect should be estimated at two or possibly more levels of the intrinsic variable.

Illustration. In a study of the relation between child anxiety and variables measuring inconsistent behaviour of the mother and support provided by the father, it was necessary to include an interaction term between the two explanatory variables. One interpretation was found by median dichotomizing on the support provided by the father; the correlation between manifestation of anxiety in the child and inconsistent behaviour in the mother was appreciably higher when the support of the father was low than when it was high.   □

Finally if the regression coefficient under study is that of an intrinsic variable, the inclusion of the variable will normally be intended to improve the relevance of the regression coefficients on treatment variables by adjusting for lack of balance in comparisons, and therefore detailed interpretation of the regression coefficient on the intrinsic variable will not normally be considered.

Illustration. A common application of multiple regression is to the adjustment of comparisons for lack of balance with respect to confounding variables. Suppose, for example, that x_1 takes the value 1 for vegetarians and 0 for nonvegetarians and it is required to assess the effect of vegetarian life-style on, say, incidence of high blood pressure or other health outcome from a cross-sectional study with a number of socio-medical measurements on each individual. Adjustment of the crude difference between vegetarians and non-vegetarians is essential if the two groups differ in age distribution, gender distribution or socio-economic class. One way to do this is by defining suitable variables $x_2, \ldots, x_q$ for entry into a multiple regression equation.

On the other hand, adjustment for cholesterol level would not be right as far as the primary comparison is concerned. Differences in cholesterol level might well be a consequence of x_1; inclusion of it would, however, be needed to address the question of whether any differences associated with x_1 are explained by cholesterol. Such issues are clarified by the graphical representations we shall discuss later.   □

Illustration. In yet other cases adjustments are formally possible but substantively meaningless. For example, suppose that the differences between men and women in some variable y are of interest and that x is height. To adjust for the differences in height between men and women would be to set up a nonsensical comparison in effect of very short men with very tall women. In any case gender is normally an intrinsic rather than a treatment variable so that direct comparisons of men with women will rarely be relevant. □

2.8 Nonspecific variables and the ecological issue

We now deal briefly with the role of nonspecific variables, i.e. variables representing broad groupings of individuals not strongly characterized by particular properties. In the discussion it is important to identify what constitutes an individual and this can be different for different aspects of an analysis.

Illustration. In an educational study data might be obtained on children, grouped by class within a school, by schools within a geographical area and by geographical areas within countries. For different purposes, individuals might be children or schools or geographical areas. Schools, geographical areas and countries would normally best be regarded as nonspecific variables.   □

First, the relation between a particular response variable and a particular explanatory variable may be different at different hierarchical levels. Figure 2.13 gives a schematic representation of this, groups in fact being a special instance of a moderating variable. The relation between y and x is quite different between and within groups; a single straight-line relation fitted ignoring groups would be closer to the between-group regression than to the within-group regression.

If data are available only at the level of group means, the interpretation of a relationship found between y and x as if it applied to individuals within a group is in general misleading, possibly seriously so. Such a misinterpretation is sometimes called an ecological fallacy.

A broad guide is that the most insightful relations are likely to be found at the lowest hierarchical level at which individuals can be treated as independent.

Illustration. In nutritional epidemiology it may be desired to study the relation between some aspects of diet and long-term ad-

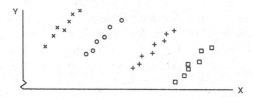

Figure 2.13 *Schematic representation of possible deviation of overall and within-group regressions.*
Positive slopes within each group; different symbols indicate different groups; overall regression is close to that through between-group means, has negative slope.

verse events, such as death from heart disease. Ideally this should be done at the level of individual persons. Examination of, say, national disease rates versus gross consumptions of materials of various kinds, i.e. an analysis treating countries as individuals, would be subject to a potential ecological fallacy. To put the point slightly differently, countries being a nonspecific variable, there are many features differentiating countries that could distort the relation. However, measurement of long-run food consumption of individual people is extremely difficult and subject to large errors of measurement, whereas the country-based data are relatively reliable. In fact a number of important conclusions have been reached by country-based or large-region-based studies.   □

On the whole, elaborate study of the pattern of variation at a more detailed hierarchical level than the one of primary interest is usually justified only to find summary measures on which to base the main analysis.

Illustrations. In an educational study in which analysis of inter-school differences is the primary objective, detailed analysis of individual children and classes of children would be required only to find summary measures for each school on which to base the primary comparisons. In a different context, if a time series of observations is obtained on each individual under study, detailed analysis of these time series would be required only to derive summary measures and possibly some indication of their precision.   □

Where the main object is the study of differences between levels of a nonspecific variable, such as differences between countries, it will be desirable to identify more specific features of, for example, countries which appear to explain relations that are present.

Illustration. In sociology it has been noted in connection with Durkheim's studies of inter-country differences in suicide rate that it is meaningless to regard the names of countries as providing an explanation of the differences found and in such comparisons an important conceptual problem is involved in identifying suitable explanatory variables. □

A somewhat related notion is the use of aggregated data, where the aggregation may be over individuals at each of a number of time points, or over time for each of a number of individuals. Such aggregation can appreciably simplify analysis, bypassing the need for detailed examination of some aspects of the variation, but care is needed in relating the properties of the aggregated data back to the behaviour of individuals.

Illustration. Suppose that in analysing psychological test scores concerning the attitude of diabetic patients to their disease, blood sugar levels are available weekly for 10–20 weeks, measured under standardized conditions. The nature of the variation in time of blood sugar level might well be of intrinsic interest, but as an explanatory variable for the relation with test score it may be best to replace each series of values by the average. If variability within patients is considered potentially explanatory, the average can be supplemented by further summary statistics, such as the standard deviation regarded as an individual explanatory variable measuring the property that might be called variability of blood pressure. □

2.9 Complex nature of variability

Somewhat connected with the discussion in the previous section is the recognition that irregular variation often has a quite complex structure and is not totally random.

Illustration. In measuring the blood pressure of a particular individual, variation can occur associated with the observer, with the particular measuring instrument used, with the behaviour of the patient in the period before the measurement is taken and with the presence of alarming individuals such as members of the medical profession; repeated measurements taken minutes, hours, days,... apart will vary, even for an individual who is in some sense in a stable state medically. The relevance of these different sources of variability will depend on the objectives of the study but some appreciation of the relative importance of the different sources can be important for sensible study design.   □

Similar principles apply quite generally. For instruments such as questionnaires to assess states or attitudes of mind the need for standardization of the measurement environment and procedure is particularly critical; in the physical sciences standardization of measurement has been finely tuned.

The consequences of uncontrolled variability on overall comparisons, including regression coefficients, are reduced by replication, i.e. by increasing the number of independent individuals in the study.

2.10 Examination of the adequacy of models

It is implicit in the above discussion that an essential part of careful statistical analysis is the examination of model adequacy and, where necessary, the development of improved models.

Three phases have to be considered concerning respectively (i) the substantive focus of the investigation, (ii) the translation of this into the primary features of a model, and finally (iii) the formulation of secondary features.

The isolation via preliminary analysis of new and unanticipated substantive questions or hypotheses can be of great value but the assessment of uncertainty then raises major conceptual difficulties. Nominal levels of statistical significance can be calculated in whatever standard way may seem relevant, but it is clear that the real uncertainty is greater and may be very much greater than is indicated by such a nominal level of significance. An allowance for selection of an effect as the most striking of many that were or might have been studied is required, but the process of selection will often be ill defined.

In fields where studies can be repeated reasonably cheaply and quickly it is reasonable to insist that such findings should be confirmed by independent study, i.e. by replication or more commonly by including in a new study design tests of the original findings. Often, however, this is feasible only in the long term and in such cases it is desirable to label such findings very clearly as hypothesis-generating rather than as directly confirmed by the study under immediate analysis.

Illustration. In comparisons of two or more medical treatments the main objective may be clear, the comparison of, say, survival in two or more groups. There is, however, always the possibility of treatment by intrinsic variable interaction, i.e. that the treatment effect is not the same for all individuals. If this is examined by looking at interaction with a few selected features there is no special problem, and indeed such analysis is usually desirable, but if a free-ranging search is made through many individual features or combinations of features looking for interactions then assessment of realistic significance becomes impossible.

The difficulty is even more severe if, instead of looking at interactions, a search is made for subgroups of patients showing a substantial difference between treatments. Thus one might find a nominally highly significant treatment difference for male smokers in a particular age band with no evidence of interesting differences for other groups of individuals. An allowance for multiple testing would need a specification of the subgroups of individuals who had or might have been examined and this is usually not feasible.

These issues arise quite commonly, especially in fields where major systematic effects are relatively rare. Practical experience in large clinical trials for detecting relatively small effects seems to suggest that interactive effects are often artefacts; on the other hand, it would be naive to assume that patients can be diagnosed so selectively that the same treatment is appropriate for all patients in a particular category.   □

The second and third types of issue of model adequacy do not raise such difficult matters. If the detailed form of the model comes from substantive considerations, it will be of intrinsic interest to examine the adequacy of fit. If the form is chosen totally empirically to fit the data, the issue may be less critical, a key requirement being to use a common form of model as far as feasible to permit comparison between related sets of data. For secondary aspects the main question is whether any departure from the model is sufficiently large to invalidate the interpretation of the data either by the introduction of bias or by degradation of efficiency.

Illustrations. If substantive considerations suggest a Michaelis–Menten equation, departure from it would often be of subject-matter interest. If substantive considerations suggest a smooth monotonic relation between two variables, assumed provisionally to be linear, a small departure from linearity would often not be of major direct subject-matter concern unless it represented a non-monotonic relation. The main requirement would be to find a representation that will provide a reasonable fit to bodies of similar data, so that comparison of the sets of data can be encapsulated in a small number of parameters.   □

Techniques for examining model adequacy are of a number of types. The least formal is to contrive a plot, or less commonly a tabular arrangement, showing under the model to be assessed either a random scatter or a straight-line display. We shall encounter a number of examples later.

The most satisfactory formal procedure is to enlarge the original model in a meaningful way by including additional parameters vanishing under the original model and to apply formal statistical arguments for assessing the values of the additional parameters. Such procedures do, however, need supplementing at least with some qualitative inspection, because it is clearly possible that the original and enlarged model are very close to each other and that both are very bad fits to the data.

A related but less satisfactory procedure is to compute one or more test statistics whose distribution under the original model is known and to compare the observed value of these statistics with that known distribution, regarding extreme values as evidence of departure from the original model. A disadvantage of such a procedure is that it has no direct diagnostic power, i.e. a significant departure may give no immediate indication of how to improve the original model.

Illustrations. In the simple linear regression model where interest lies in testing linearity of the regression function, direct plotting of the data or of residuals, defined as differences between the observed values of Y and fitted values, would be a direct graphical test of linearity.

The simplest formal test is based on elaborating the model, usually by adding a quadratic term so that

$$Y_i = \alpha + \beta(x_i - a) + \gamma(x_i - a)^2 + \epsilon_i.$$

Here a is a convenient reference level, often but not necessarily the mean value of x in the data, used primarily to avoid numerical instability in the fitting.

Other functions than the square can be used if they are more likely to represent a particular kind of departure of interest. There can be two rather different attitudes to such a model. It may be regarded as a serious candidate for the representation of the data if linearity fails. Alternatively, the squared term may be regarded just as a convenient device for detecting departure from linearity, any departure so detected then being explored in more detail.

A formal test for significant overdispersion is obtained by grouping the data into sets with equal or nearly equal values of X and then comparing the dispersion of the group means of Y around the fitted line with the dispersion of Y within groups. A significant overdispersion of the group means does not have an immediate interpretation in terms of the shape of the departure from linearity so that in the presence of smooth nonlinearity the test has poor diagnostic properties.

Other aspects of the model, such as the constancy of variance and the normality of the error distribution, can be tested by methods of the three different types although the details are a little more involved.   □

A clear distinction has to be drawn between the statistical significance of a lack of fit and its subject-matter importance. With small amounts of data quite large discrepancies may well be within the limits of random error. With large amounts of data a significance test may detect lack of fit that is, however, of no substantive importance. In the latter case a subject-matter judgement is necessary on how to proceed. Models are inevitably idealized. Thus departures from the model are inescapable and with sufficiently large amounts of data are likely to be detectable; it is important that attention is focused on departures of substantive importance. The principle that lack of fit detected in the primary aspects of the model must be reported is, however, very important as some protection against arbitrariness.

2.11 Some complications

Any serious analysis of data, especially if the data are at all extensive, must, as already noted in Chapter 1, examine data quality and, of course, achievement of this is a key issue at the design stage of any investigation. Among the matters to be considered are bias, i.e. systematic error, missing values and outliers.

Since most studies are in some sense comparative, a constant systematic error in one variable affecting all the individuals in a study would for many purposes be unimportant. Selective bias affecting only some groups of individuals may entirely distort conclusions. Its avoidance is largely an issue of design; whenever possible, blind checks on comparability should be included.

The analysis of missing data raises technical issues of analysis that we shall not discuss in detail. If the proportion of missing values is small and they are missing at random, i.e. for reasons not selectively concerned with the value of the variable in question, then the simplest procedure is to impute notional values for the missing data either by some simple averaging procedure or via the regression of the missing variable on the variables that are recorded for the individual in question. If the proportion of missing values is at all appreciable some allowance is needed for the degradation of precision resulting, or it may be decided to get better data. For further discussion, see Section 8.6.

Outliers and more broadly anomalous values or small groups of values arise very commonly, and inspection of the data for their presence is essential. This raises difficulties with very large sets of data. We shall return to the methods involved later. In general, however, it is dangerous simply to reject or automatically downweight and then to forget about such anomalous values. They may indeed be the result of recording errors or other gross mistakes, but there is sometimes also the possibility that they are a warning of important unanticipated effects. It is frequently desirable to repeat analyses with and without suspect values as a check on their influence on the conclusions.

Outliers can be detected from logical inconsistencies in data, from values extreme in the marginal distribution of the variable in question, and from individuals who break an otherwise strong dependence between variables.

Illustrations. Data in which the unit of study is the family might include numbers of males and females and perhaps separately numbers of individuals in different age ranges. If the totals do not agree there is a recording error which, while not necessarily an outlier in being extreme, should be corrected before the start of formal analysis.   □

Figure 2.14 *Example with outlier detection possible from bivariate but not from univariate distributions.*

While an observation on a variable can be rejected if it lies far outside the effective range of variation when that is known from prior grounds, the more sensitive detection obtained via correlated pairs of variables, for example log height and log weight, is illustrated in Figure 2.14. The anomalous individual is not extreme via either variable on its own but stands out clearly in the bivariate distribution.

2.12 Causality

In the formal theory of statistical analysis, conclusions are expressed via probability models and confidence limits for parameters of interest. To be useful such statements have then to be interpreted in subject-matter terms and in particular related to background knowledge of the subject, previous similar work and so on.

As has been stressed in previous sections, models should be chosen as far as feasible to aid this process of interpretation.

An ideal interpretation would normally be via an understanding of the process that generated the data in terms of concepts at some 'deeper' level. This could be called a causal explanation.

Illustration. It might be possible to establish an interpretation of the effect on patients of a drug, as observed in a clinical trial, via a pharmaco-kinetic model of the action of the drug, or to interpret the effect on health status or quality of life via the effect on blood pressure and biochemical measurements, although this would raise deeper issues of interpretation.   □

It is rarely likely to be possible to draw firm conclusions of this kind from a single study or even from a series of studies of similar structure. We shall, therefore, in this book for the most part

avoid the use of the word *causal* on the grounds that no statistical
analysis of the kind we consider could on its own provide convincing evidence of causality in the strong sense just mentioned.
Our formulations are, however, often designed to point towards explanations that are potentially causal; this is largely why we put
considerable emphasis on the distinction between response, intermediate response and explanatory variables.

The word *causal* is sometimes used in one of two weaker senses.
The first is that of a statistical association such that all reasonable regression equations for predicting the response contain the
explanatory variable in question, i.e. the apparent dependence cannot be explained away via other explanatory variables referring to
time points prior to the explanatory variable in question. This is
essentially one possible definition of causality; its implementation
will always be partial because the number of alternative explanatory variables that can be examined will be limited.

The other definition is that there is convincing evidence that if
the explanatory variable, a treatment variable, were to be changed
for an individual, other things remaining unchanged, then the response on the individual in question would be systematically al-
tered.

These are important ideas which will be discussed further in
Section 8.7.

2.13 Bibliographic notes

Section 2.1. The connection between substantive research hypotheses and graphical models is stressed by Wermuth and Lauritzen
(1990).

Sections 2.2–2.4. The use of graphical representations to describe
complex dependencies goes back to path analysis introduced by
the geneticist Wright (1921). Applications of the ideas to sociology
and to econometrics are discussed respectively by Duncan (1975)
and Wold (1954). Much emphasis in this work is on partitioning
dependency into direct and indirect components.

Books emphasizing graphs as representing conditional independence are by Whittaker (1990), Edwards (1995) and Lauritzen
(1996). A general discussion of conditional independence was given
by Dawid (1979) who introduced the notation for it now widely

used. For a review of the connections with spatial processes, see Isham (1981).

The independence interpretation of special log-linear models for contingency tables and of Dempster's (1972) covariance selection models was shown by Wermuth (1976). A characterization and further study of these models which can be represented by full-edge undirected graphs was given for contingency tables by Darroch, Lauritzen and Speed (1980), building on earlier work by Goodman (1970; 1971), and for Gaussian systems by Speed and Kiiveri (1986). For tests associated with a special class of independence models in the multivariate normal distribution see Andersson and Perlman (1995).

The possibility of reinterpreting models represented by undirected full-line graphs as dependence models in fully directed graphs was obtained by Wermuth (1980) for Gaussian systems, by Wermuth and Lauritzen (1983) for contingency tables and by Frydenberg (1990a) for conditional Gaussian systems. Models corresponding to full-edge joint-response chain graphs were introduced by Lauritzen and Wermuth (1989) and studied in detail by Frydenberg (1990b). A survey of results was given by Lauritzen (1989).

Separation criteria were given for fully directed acyclic graphs by Pearl (1988) and Lauritzen et al. (1990), for full-edge joint-response chain graphs by Frydenberg (1990b) and for dashed-edge undirected graphs by Kauermann (1996). Properties of general lists of independence statements and of corresponding probability models have been given by Geiger and Pearl (1993).

Graphs with two types of edge were introduced by Cox and Wermuth (1993a) who emphasized the connection with various special kinds of statistical model. For a review of the application of directed acyclic graphs to probabilistic expert systems, see Spiegelhalter et al. (1993).

The construction of covariance and concentration graphs from a given directed acyclic graph was described by Pearl and Wermuth (1994). Efficient algorithms based on so-called prime graphs were given by Leimer (1993) for orienting an undirected full-line graph to become directed acyclic such that its independence structure is preserved or it is decided that there is no such solution.

Section 2.5. See the references for Section 8.3.

Section 2.7. For a discussion of moderating variables, see Wermuth

(1987) and Geng and Asano (1993). For the Yule–Simpson paradox, see Yule (1900), Simpson (1951) and Snedecor and Cochran
(1967, p. 472).

Section 2.8. The ecological issue is discussed in econometrics (Stoker,
1986) under the term 'aggregation', in the sociological literature
and in epidemiology, especially nutritional epidemiology. See Plummer and Clayton (1995).

Section 2.9. The notion of hierarchies of error has a long history in
the context of balanced experimental design. For formal methods
for general unbalanced data, see Goldstein (1995).

Section 2.12. For references on causality, see the notes for Section
8.7.

CHAPTER 3

Technical considerations

3.1 Preliminaries

We now turn to detailed issues bearing first on model formulation and then, in Chapter 4, on fitting and estimation. For quantitative variables a central role is played by means and covariances and less commonly by higher-order moments. The first two give the basis for a summary of associations when interrelationships between variables are essentially linear and a complete summary of these associations when the special form of joint distribution, the multivariate normal distribution, holds.

We therefore give in Sections 3.2–3.5 a fairly full account of covariance matrices, the closely related topic of least-squares linear regression relations and introduce some further conventions for the graphical representation of such special situations. The emphasis initially is on the formulation of models; the implications for statistical analysis are developed in Chapter 4.

A much briefer account of the corresponding results for binary and mixed variables is given in Sections 3.7 and 3.10.

Some familiarity with matrix notation is assumed in much of the discussion.

3.2 Covariance matrix

3.2.1 Some definitions

Suppose that on each individual p continuous variables are measured. It is convenient when we are dealing with vector random variables to regard them as column vectors, although this leads to some minor notational problems when we consider sets of individuals. That is, we consider the $p \times 1$ column vector Y which we write conventionally as $(Y_1, \ldots, Y_p)$. When there are n individuals each with a value of Y we write the full set of random variables again as Y, with elements $Y_{is}(i = 1, \ldots, n; s = 1, \ldots, p)$, this time an $n \times p$

matrix with rows indexing individuals and columns indexing the
component variables. That is, each individual contributes a row.
This apparent inconsistency of notation does, in the end, lead to
somewhat simpler formulae.

We describe the distribution of a single column vector Y by
defining the mean vector and covariance matrix

$$E(Y) = \mu = (\mu_1, \ldots, \mu_p),$$

$$\text{cov}(Y) = E\{(Y - \mu)(Y - \mu)^T\} = \Sigma,$$

say, the covariance matrix Σ being a $p \times p$ symmetric matrix with
(r, s) element

$$\text{cov}(Y_r, Y_s) = \sigma_{rs} = E\{(Y_r - \mu_r)(Y_s - \mu_s)\}$$

and the superscript T denoting a transpose. Sometimes we write
Σ_{YY} to show the random variable involved.

If $r = s$,

$$\text{cov}(Y_r, Y_r) = \text{var}(Y_r) = \sigma_{rr} = \sigma_r^2,$$

the variance of Y_r.

If there are a small number of components, we use a slightly
different notation, writing if, for example, there are only two components $Y = (U, V)$,

$$E(Y) = (\mu_U, \mu_V),$$

$$\sigma_{UV} = \sigma_{VU} = \text{cov}(U, V) = E\{(U - \mu_U)(V - \mu_V)\},$$

$$\sigma_{UU} = \text{var}(U).$$

The covariance has the dimensions of the product of the two
random variables involved. To interpret the strength of a linear
relation, it is sometimes convenient to take a dimensionless form
obtained in effect by standardizing variables to have unit variance
and thus to define the correlation coefficient between Y_r and Y_s to
be

$$\text{corr}(Y_r, Y_s) = \sigma_{rs}/(\sigma_{rr}\sigma_{ss})^{1/2} = \rho_{rs},$$

say.

The corresponding matrix is called the correlation matrix. It
has unit diagonal elements.

Among the rules for manipulating covariances note in particular
that if a new $m \times 1$ vector of random variables Z is produced from
Y by linear transformation,

$$Z = AY,$$

where A is an $m \times p$ matrix of constants, then the covariance matrix of Z is

$$\Sigma_{ZZ} = A\Sigma_{YY}A^T.$$

In particular if A is a row vector and Z is a scalar, $Z = \Sigma a_r Y_r$, say, then

$$\text{var}(Z) = \Sigma a_r a_s \sigma_{rs}.$$

Because variances are nonnegative and zero if and only if the random variable concerned is constant, it follows that any covariance matrix is positive semi-definite and positive definite whenever there is no linear combination of the components that is constant with probability one.

In particular, if we apply this argument to a pair of random variables it follows that

$$-1 \leq \rho_{rs} \leq 1,$$

with equality if and only if Y_r and Y_s are deterministically linearly related, a case that will not arise directly in empirical studies except via the unwitting use of component variables constrained by definition, for example to sum to 1 or to the value of one of the other component variables.

3.2.2 Linear least-squares regression

We begin by considering a pair of random variables (U, V). If U is independent of V then by the product property of expectations $\text{cov}(U, V) = \sigma_{UV} = 0$. While $\sigma_{UV} = 0$ does not imply the much stronger statement of independence, i.e. that any probability statement about V is unchanged by being given the value of U, vanishing of the covariance does imply that in a sense there is no systematic linear relation between V and U. We see this as follows.

Suppose that we find an approximation to the conditional expectation $E(V \mid U)$ by a general linear expression in U, which can be written in the form

$$E(V \mid U) = \tilde{V} = \mu_V + \alpha + \beta(U - \mu_U),$$

where (α, β) are constants to be chosen. Then a direct calculation shows that the mean squared error is

$$E(V - \tilde{V})^2 = \sigma_{VV} + \beta^2 \sigma_{UU} - 2\beta\sigma_{UV} + \alpha^2,$$

and this is minimized by the choices

$$\alpha = 0, \ \beta = \sigma_{VU}\sigma_{UU}^{-1},$$

say. Further, the resulting minimum mean squared error is

$$\sigma_{VV} - \sigma_{VU}\sigma_{UU}^{-1}\sigma_{UV} = \sigma_{VV.U} = \sigma_{VV}(1 - \rho_{UV}^2),$$

say, where the reason for writing the left-hand side in the unreduced form shown is to establish a connection with the general vector case.

We call β the least-squares regression coefficient of V on U and $\sigma_{VV.U}$ the residual variance or conditional variance of V given U. These results and definitions apply to any distribution but their relevance is small if the relation between $E(V \mid U)$ and U is very nonlinear. Quite often we want to apply linear regression ideas when the linear least-squares regression accounts for most but not all of the relation between $E(V \mid U)$ and U and we then call the relation *quasi-linear*.

The results generalize when there are more than two variables involved.

First we consider a single component V to be related to a set of components $(U_1, \ldots, U_q)$. A generalization of the argument above shows that the linear minimum mean squared error predictor of V is

$$\mu_V + \Sigma\beta_{Vs.\backslash s}(U_s - \mu_s),$$

where the $1 \times q$ vector $\beta_{V|U}$ of regression coefficients $\beta_{Vs.\backslash s}$, specifying the regression of V on U_s given the remaining components, is

$$\beta_{V|U} = \Sigma_{VU}\Sigma_{UU}^{-1}$$

where Σ_{VU} is the row vector with elements σ_{VU_s}. The mean squared error about the regression is

$$\sigma_{VV.U} = \sigma_{VV} - \Sigma_{VU}\Sigma_{UU}^{-1}\Sigma_{UV},$$

where $\Sigma_{VU} = \Sigma_{UV}^T$.

For V consisting of several components it is convenient to change notation, to partition Y into two components (Y_a, Y_b) of dimensions $p_a \times 1, p_b \times 1$, respectively and to consider the regression of Y_a component by component on Y_b. The regression coefficients of each component are given by the formulae above for V regressed on U and we may assemble the regression coefficients into a $p_a \times p_b$

matrix $B_{a|b}$ given by

$$B_{a|b} = \Sigma_{ab}\Sigma_{bb}^{-1}.$$

In this each row corresponds to a component response variable.

The covariance matrix of deviations from the regression is

$$\Sigma_{aa.b} = \Sigma_{aa} - \Sigma_{ab}\Sigma_{bb}^{-1}\Sigma_{ba}.$$

Here, the covariance matrix Σ of Y, and the *concentration matrix*, Σ^{-1}, have been partitioned corresponding to the partition of Y, to give

$$\Sigma = \begin{pmatrix} \Sigma_{aa} & \Sigma_{ab} \\ \Sigma_{ba} & \Sigma_{bb} \end{pmatrix}, \quad \Sigma^{-1} = \begin{pmatrix} \Sigma^{aa} & \Sigma^{ab} \\ \Sigma^{ba} & \Sigma^{bb} \end{pmatrix},$$

where the partitioning can be generalized when there are more than two components.

A direct approach to these results is as follows. We consider the pair of random variables

$$Y_{a.b} = Y_a - B_{a|b}Y_b, \quad Y_b.$$

These two random variables are required to be uncorrelated; the first is the vector of deviations from the least-squares regression. On postmultiplying the first component by Y_b^T and taking expectations we obtain the equation defining $B_{a|b}$, whereas forming the covariance matrix of the first component we obtain $\Sigma_{aa.b}$. Among the consequences of these formulae are that the inverse covariance matrix, or what we call the concentration matrix, of $Y_{a.b}$ is the upper left-hand section of Σ^{-1} when partitioned, i.e. that

$$\Sigma_{aa.b}^{-1} = \Sigma^{aa},$$

and that the covariance matrix of Y_b, namely Σ_{bb}, has concentration matrix

$$\Sigma_{bb}^{-1} = \Sigma^{bb.a} = \Sigma^{bb} - \Sigma^{ba}(\Sigma^{aa})^{-1}\Sigma^{ab}.$$

To generalize the argument when there are more than two component subvectors we proceed as follows.

With three subvectors, we define $Y_{a.bc}$ as the residual of Y_a from its regression on (Y_b, Y_c), writing

$$Y_{a.bc} = Y_a - B_{a|b.c}Y_b - B_{a|c.b}Y_c.$$

The full matrix of regression coefficients of Y_a on (Y_b, Y_c) is defined via

$$B_{a|bc} = (B_{a|b.c}, B_{a|c.b}) = (\Sigma_{ab}\Sigma_{ac}) \begin{pmatrix} \Sigma_{bb} & \Sigma_{bc} \\ \Sigma_{cb} & \Sigma_{cc} \end{pmatrix}^{-1},$$

where the matrices on the right-hand side refer to the variable (Y_b, Y_c) and thus can be partitioned into contributions from the separate components.

We can thus write, recalling that the vectors are column vectors,

$$(Y_{a.bc}, Y_{b.c}, Y_c) = A(Y_a, Y_b, Y_c),$$

where A is an upper-triangular block matrix, with identity blocks down the diagonals and matrices of minus the regression coefficients in the off-diagonal blocks. The inverse of A has the same form, with the signs of the off-diagonal blocks changed but with the interpretation as regression coefficients in one regression equation lost. By construction the covariance matrix of the left-hand side is block-diagonal,

$$D = \mathrm{diag}(\Sigma_{aa.bc}, \Sigma_{bb.c}, \Sigma_{cc}).$$

It follows that the concentration matrix has a block-triangular decomposition as

$$\Sigma^{-1} = A^T D^{-1} A.$$

If the vector Y is partitioned into single components, we write correspondingly the column vector of residuals as

$$(Y_{1.2,\dots,p}, \dots, Y_{p-1.p}, Y_p)$$

and

$$D = \mathrm{diag}(\sigma_{11.2,\dots,p}, \dots, \sigma_{p-1,p-1.p}, \sigma_{pp}).$$

It is important that the linear least-squares regression exists for any random variables subject only to the technical condition of finite variance, this condition ruling out distributions that are so long-tailed that squared error is inapplicable as a measure of discrepancy. The least-squares regression has errors that are uncorrelated with the explanatory variables. This is a very much weaker condition than having the errors independent of the explanatory variables, which would employ the strong assumption that the conditional mean of response varies linearly with the explanatory variables. The linear least-squares regression will be of some substantive interest, however, whenever the relation between the variables has at least a strong linear component, i.e. in what we termed the quasi-linear case.

3.3 Multivariate normal distribution

While the above discussion holds for any distribution of the p-dimensional random variable Y it is, as stressed above, statistically relevant only when the interrelationships of the components are predominantly linear. An important special distribution where this is so and where the results of Section 3.2.2 play a central role is the multivariate normal distribution.

The $p \times 1$ random variable Y has a multivariate normal distribution of mean μ and covariance matrix Σ, denoted by $N_p(\mu, \Sigma)$, if its probability density has the form

$$f_Y(y) = (2\pi)^{-p/2}\{\det(\Sigma)\}^{-1/2}\exp\{-(y-\mu)^T\Sigma^{-1}(y-\mu)/2\}.$$

The terminology implies that $E(Y) = \mu$ and this is seen by noting that $Y - \mu$ and $-(Y - \mu)$ have the same distribution. To prove that $\mathrm{cov}(Y) = \Sigma$ the most direct approach is to note from the block-triangular decomposition of Σ^{-1} that we can write

$$\Sigma^{-1} = CC^T,$$

where C is nonsingular, so that also

$$\Sigma = (C^T)^{-1}C^{-1}.$$

Application of the formula for transforming probability density functions under transformations of variables shows that the components of $Z = C^T(Y - \mu)$ are independently normally distributed with zero mean and variance one and hence have covariance matrix I, the identity matrix. It then follows without detailed calculation that the covariance matrix of Y is $(C^T)^{-1}C^{-1}$, as required.

When Y is partitioned as $Y = (Y_a, Y_b)$, the marginal distribution of, say, Y_b is obtained by integration over the other variable and this yields the exponential of a quadratic function of the components y_b and hence represents a multivariate normal distribution. Its mean and covariance are, by virtue of the definitions of expectation and covariance, μ_b and Σ_{bb}. That is, $f_{Y_b}(y_b)$ has the form $N_{p_b}(\mu_b, \Sigma_{bb})$. That is, in the multivariate normal distribution all marginal distributions are normal.

The conditional density of Y_a given $Y_b = y_b$ is the ratio

$$f_Y(y)/f_{Y_b}(y_b).$$

Inspection of the exponential argument shows it to be quadratic in y_a and therefore representing a multivariate normal distribution.

Moreover the coefficient of the quadratic term in y_a is constant independent of y_b and the coefficient of the linear term is linear in y_b. Therefore the conditional covariance matrix is constant and the conditional mean is linear in y_b. These must coincide with results of Section 3.2. That is, in the multivariate normal distribution all conditional distributions are multivariate normal.

The key formulae are that

$$E(Y_a \mid Y_b = y_b) = \mu_a + \mathrm{B}_{a|b}(y_b - \mu_b),$$
$$\mathrm{cov}(Y_a \mid Y_b = y_b) = \Sigma_{aa.b} = \Sigma_{aa} - \Sigma_{ab}\Sigma_{bb}^{-1}\Sigma_{ba}.$$

For most of our discussion the detailed properties of the multivariate normal distribution are not needed; its importance lies in indicating one set of circumstances in which the linear least-squares regression equation and its associated properties have a strong interpretation and justification.

3.4 Concentration matrix

A key role in the definition of the multivariate normal distribution is played by the inverse covariance matrix, Σ^{-1}, which we have called the concentration matrix, and which is sometimes referred to in the literature as the precision matrix.

We recall that its elements have a direct interpretation and that whenever the random variable Y and covariance matrix Σ are partitioned, we partition the concentration matrix conformally, writing, for example, the concentration matrix for two components as

$$\begin{pmatrix} \Sigma^{aa} & \Sigma^{ab} \\ \Sigma^{ba} & \Sigma^{bb} \end{pmatrix}.$$

We recall also that by the arguments of Section 3.2 the inverse of sections of the concentration matrix determines a conditional covariance matrix. Therefore, if a particular element of the concentration matrix vanishes, say $\sigma^{rs} = 0$, then the components Y_r and Y_s are uncorrelated after taking residuals from linear regression on all the other components. In the multivariate normal distribution, Y_r is conditionally independent of Y_s given all the remaining variables.

More explicitly the concentration matrix, being a symmetric positive definite matrix can be regarded as the covariance matrix of a new vector of random variables, and in fact if $Z = \Sigma^{-1}Y$ we have that the covariance matrix of Z is, assuming zero means for simplicity,

$$E(ZZ^T) = \Sigma^{-1}\Sigma\Sigma^{-1} = \Sigma^{-1}.$$

Further,

$$E(YZ^T) = E(YY^T\Sigma^{-1}) = I,$$

the identity matrix, implying in particular that each component of Z is uncorrelated with all the other nonmatching components of Y. This leads to a direct interpretation of the vector Z as having components proportional to the complete residuals, i.e. to the difference of, say, Y_j from its regression on all other components of Y. For we can write, for example,

$$Z_1 = \sigma^{11}Y_1 + \ldots + \sigma^{1p}Y_p,$$

$$Y_1 = -Y_2\sigma^{12}/\sigma^{11} - \ldots - Y_p\sigma^{1p}/\sigma^{11} + Z_1/\sigma^{11},$$

and the last term on the right-hand side is uncorrelated with the other terms. Therefore the equation represents a least-squares regression equation. It follows in general that

$$\beta_{rs.\backslash rs} = -\sigma^{rs}/\sigma^{rr},$$

where the regression coefficient on the left-hand side is that of Y_r on Y_s given the remaining components. If we convert the regression coefficient into a correlation, for example via the product with $\beta_{sr.\backslash sr}$, we have that the partial correlation between Y_r, Y_s eliminating regression on the other components is

$$\rho_{rs.\{1,\ldots,q\}\backslash\{rs\}} = -\sigma^{rs}/(\sigma^{rr}\sigma^{ss})^{1/2}.$$

In particular for just three variables we have that

$$\rho_{12.3} = -\sigma^{12}/(\sigma^{11}\sigma^{22})^{1/2} = \frac{\rho_{12} - \rho_{13}\rho_{23}}{[(1-\rho_{13}^2)(1-\rho_{23}^2)]^{1/2}}.$$

This can be derived also as the correlation coefficient between $Y_1 - \beta_{13}Y_3$ and $Y_2 - \beta_{23}Y_3$. The corresponding recursive formula for regression coefficients is usually written in the form

$$\beta_{12} = \beta_{12.3} + \beta_{13.2}\beta_{23}.$$

Here β_{12} is the regression coefficient of Y_1 on Y_2 ignoring Y_3, i.e. marginalizing over the distribution of Y_3, whereas $\beta_{12.3}$ is the regression coefficient of Y_1 on Y_2 in the multiple regression equation of Y_1 on Y_2, Y_3. As noted in Sections 1.5 and 2.7, β_{12} and $\beta_{12.3}$ are capable of having opposite signs although this requires a relatively large value of β_{23} of appropriate sign, i.e. for the two explanatory variables concerned to be highly correlated.

Table 3.1 *Correlations among four psychological variables for 684 stu-dents. Marginal correlations (lower triangle), partial correlations given the other two variables (upper triangle).*

	Y	X	V	U
Y, state anxiety	1	0.45	0.47	−0.04
X, state anger	0.61	1	0.03	0.32
V, trait anxiety	0.62	0.47	1	0.32
U, trait anger	0.39	0.50	0.49	1

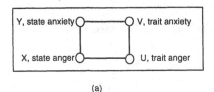

(a)

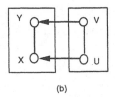

(b)

Figure 3.1 *Independence graphs with full-edge 4-cycles, both correspond-ing to data in Table 3.1.*
(a) Concentration graph capturing zero pattern in partial correlations.
(b) Independence-equivalent joint-response chain graph with trait vari-ables treated as explanatory to states, Y regressed on X, V, U and X regressed on Y, V, U. In both graphs $Y \perp\!\!\!\perp U | (X,V)$ and $X \perp\!\!\!\perp V | (Y,U)$. ,

It is useful in applications where a correlation matrix is exam-ined to show the corresponding partial correlations alongside the marginal correlations, for example in the upper triangle of the $p \times p$ matrix.

Illustration. In a study of anger and anxiety, test scores were obtained for trait anger, trait anxiety, state anger and state anx-iety. The trait variables are intended to give long-term features of an individual, the state variables short-term features. Table 3.1 shows the estimated correlations in the lower triangle and partial correlations in the upper triangle.

While the marginal correlations show no striking pattern, two of the partial correlations are very small. The graphical representa-tion in Figure 3.1 in the form of a chordless 4-cycle shows this; one direct reinterpretation is that to predict, for example, state anger we need state anxiety and trait anger, but not trait anxiety. We shall return to the interpretation of such a structure in Sections 5.3 and 8.5. □

3.5 Special matrices

Any positive definite matrix is a possible covariance matrix, or indeed a possible concentration matrix. To represent the variances and covariances of a set of observations on n independent individuals we thus have $\frac{1}{2}p(p+1)$ parameters. Especially for moderate or large values of p it is helpful for empirical analysis and for interpretation to have special families of covariance matrices with simple structure defined by a smaller number of unknown parameters.

Some of these special forms will emerge during the detailed discussion later in this book; here we outline a few of the simpler possibilities.

Superficially the most obvious simplification arises when a number of correlations are zero, i.e. such that the corresponding pairs of variables are, in the multivariate normal case, marginally independent. Another possibility is that elements of the concentration matrix are zero, i.e. that the corresponding pairs of variables are conditionally independent given the remaining variables. These possibilities, especially the former, are particularly useful in cases where most of the variables are independent or nearly so, with a few substantial correlations standing out. The two possibilities correspond respectively to covariance and concentration graphs with appreciable numbers of missing edges and their statistical properties are quite different. The theory of concentration graphs is much richer than that of covariance graphs and the associated theory of estimation is simpler too.

A quite different special case is the intra-class or simple variance-component model conveniently written in the form

$$Y_r = \mu_r + \sigma_\xi \xi + \sigma_\eta \eta_r,$$

where (ξ, η_r) are independent and identically distributed according to the standard normal distribution. It follows that

$$\sigma_{rr} = \sigma_\xi^2 + \sigma_\eta^2,$$

$$\sigma_{rs} = \sigma_\xi^2 \ (r \neq s).$$

Conversely if all variances are equal and all covariances equal and positive the above representation of the Y_r is possible. More general variance-component models will typically generate a block structure in the covariance matrix.

If we generalize the above model to the single-factor model of factor analysis,

$$Y_r = \mu_r + \kappa_r \sigma_\xi \xi + \sigma_\eta \eta_r,$$

where the κ_r are constants, we have that

$$\sigma_{rs} = \kappa_r \kappa_s \sigma_\xi^2 \ (r \neq s).$$

This implies the so-called tetrad condition that for unequal r, s, t, u

$$\sigma_{rs}\sigma_{tu} - \sigma_{rt}\sigma_{su} = 0.$$

Another important special class of covariance matrices arises from stationary time series.

The special forms with zero covariances between specified pairs of variables and the variance-component models are special cases of the family of linear covariance structures in which

$$\Sigma = \alpha_1 \Omega_1 + \alpha_2 \Omega_2 + \ldots + \alpha_m \Omega_m,$$

where the α_r are unknown parameters and the Ω_r are known matrices. Such representations are useful also in connection with unbalanced hierarchical error structures although we shall not develop that aspect here.

Illustration. Correlation matrices in which particular elements are constrained to be zero are a special case. Thus with four variables and two correlations constrained to zero we can take $m = 8$; all matrices have essentially only one nonzero element, the first four matrices having one in a diagonal position and the second four having in turn one in the four off-diagonal positions in which the covariance may be nonzero.

Table 3.2 gives an empirical example arising from a study of strategies to cope with stressful events. Cognitive avoidance and blunting are thought of as strategies to reduce emotional arousal, and vigilance and monitoring as strategies to reduce insecurity. Two of the marginal correlations, given below the diagonal, are very small, whereas there is no comparable simplification in the concentrations given above the diagonal. Figure 3.2 gives the corresponding graphical representation, in this case a covariance graph of dashed lines, since marginal correlations are being represented.

Table 3.2 *Correlations among four strategies to cope with stress for 72 students. Marginal correlations (lower triangle), partial correlations given the other two variables (upper triangle) .*

	Y	X	V	U
Y, cogn. avoid.	1	−0.30	0.49	0.21
X, vigilance	−0.20	1	0.21	0.51
V, blunting	0.46	0.00	1	−0.25
U, monitoring	0.01	0.47	-0.15	1

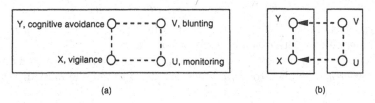

(a)   (b)

Figure 3.2 *Independence graphs with dashed-edge 4-cycles.*
(a) Covariance graph capturing zero pattern in marginal correlations of Table 3.2 and implying $Y \perp\!\!\!\perp U$ and $X \perp\!\!\!\perp V$. (b) Joint-response chain graph with Y response to V and X response to U, implying $Y \perp\!\!\!\perp U|V$ and $X \perp\!\!\!\perp V|U$; not compatible with the data in Table 3.2.

In contrast to Figure 3.1 the two graphs in Figure 3.2 are not independence-equivalent. □

3.6 Binary data

The discussion of models so far has been primarily intended for effectively continuous responses, preferably having a distribution reasonably close to multivariate normality, but failing that being such that interrelationships are predominantly linear.

We now consider other types of response variable, starting with the simplest case of a univariate binary variable, A, say. Denote the two possible values of A by 0 and 1, sometimes called for convenience failure and success; for the more specialized arguments of Section 3.7 we shall, however, denote the two values by −1 and 1.

The distribution of A is determined by

$$E(A) = P(A = 1) = \pi,$$

say; for a single observation $A^2 = A$, so that

$$E(A^2) = \pi, \ \mathrm{var}(A) = \pi(1 - \pi).$$

Thus the variance changes with π, although only rather slowly over the range $0.2 \leq \pi \leq 0.8$, say.

If we have observations on n uncorrelated individuals each giving a binary response, and all provisionally assumed to have the same value of π, then the total number of successes $S = \Sigma_j A_j$ is such that

$$E(S) = n\pi, \ \mathrm{var}(S) = n\pi(1 - \pi),$$

the distribution being of binomial form when the individuals are independent rather than merely uncorrelated.

Now suppose that we want to represent the dependence of A on a vector of explanatory variables, x. At first sight the most direct approach is to consider a linear model

$$E(A_i) = \pi_i = \Sigma_r x_{ir}\beta_r, \ E(A) = x^T\beta,$$

and to apply the method of least squares in the way to be developed below primarily for continuous responses.

There are two reasons why this may not be satisfactory, one being connected with the linking of variance and mean and the more basic one being that $0 \leq \pi_i \leq 1$, whereas the linear expression on the right is not so constrained. This may be of little concern when only values of π in the centre of the range are of interest. Even then, comparison with other sets of data with more extreme values of explanatory variables may be inhibited.

The most effective way of dealing with the constraint on π is to use a representation in which the constraint is automatically satisfied for all values of parameters and explanatory variables. For this we write

$$\Lambda(\pi_i) = \Sigma_r x_{ir}\beta_r,$$

where $\Lambda(\pi)$ is an increasing function of π chosen to go from $-\infty$ to ∞ as π goes from 0 to 1. Usually $\Lambda(\pi)$ is a known function, although in principle it could to some extent be estimated empirically given a large amount of data. The direct linear formulation can be regarded as the special case $\Lambda(\pi) = \pi - 1/2$ applying, however, only over a limited range.

There is a distinction between the more common case in which
the analysis is essentially unchanged if the roles of success and
failure are interchanged and ones where there is a fundamental
asymmetry between the two responses. In the former case the most
useful special choices for $\Lambda(\pi)$ are the logistic and probit forms with
respectively

$$\Lambda(\pi) = \log\{\pi/(1 - \pi)\},$$

and

$$\Lambda(\pi) = \Phi^{-1}(\pi),$$

where $\Phi(.)$ is the standardized normal integral, whereas in the
asymmetric case we could take

$$\Lambda(\pi) = -\log(-\log\pi),$$

or

$$\Lambda(\pi) = \log(-\log(1 - \pi)),$$

called complementary log-log laws.

These two special forms are connected theoretically with the
study of extreme values. One of the complementary log-log laws
results if the binary response is connected with an underlying unobserved continuous variable that is itself either the largest or the
smallest of a large number of independent random variables, being
one or zero depending on whether that latent variable is above or
below a threshold. Figure 3.3 shows the shapes corresponding to
linear, logistic and one of the two complementary log-log laws. The
probit form is indistinguishable from the logistic to the accuracy
used.

When a direct linear representation is used the interpretation
of a regression coefficient β is in terms of the change of probability
π per unit change in the corresponding explanatory variable x.
When the logistic version is used the interpretation is in terms of
the change in the log-odds, $\log\{\pi/(1 - \pi)\}$. When π is small this
is equivalent to proportional change in π, whereas when π is near
one the interpretation is in terms of proportional change in $1 - \pi$.

Experience in the use of logistic and other nonlinear functions
of probabilities eases the difficulties of interpreting differences and
regression coefficients calculated on the transformed scale. Even
so, presentation on a direct probability scale will often be helpful.
For example, to explain the meaning of a regression coefficient on
a continuous explanatory variable, the most direct procedure is to
fix one or more sets of reference levels of the other explanatory

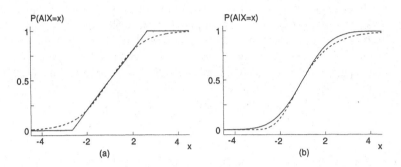

Figure 3.3 *Forms of curve for probability $P(A = 1)$ versus a single explanatory variable X.*
(a) Logistic (dashed), linear (solid); (b) log-log (dashed), cumulative normal (solid).

variables and then to plot the fitted probability from the logistic regression versus the explanatory variable, thus showing one fitted curve for each set of reference levels used. A similar method can be used for discrete explanatory variables. This is a rather general method for presenting the results of analyses on transformed scales.

This and the issue of how to present estimates representing interactions are discussed in more detail in Section 4.5 and in the specific applications of Chapter 6.

Empirical discrimination between the logistic and integrated normal models from binary data requires large numbers of observations; both approach limiting values of 0 and 1 in the tails quite rapidly and symmetrically, the logistic rather more slowly than the integrated normal. If regression coefficients are calculated from logistic, integrated normal and linear models, the last from data referring to the central range of probabilities, then to a close approximation, in a self-explanatory notation,

$$\beta_{INT} = 0.607\beta_{LOG}, \quad \beta_{LIN} = 0.144\beta_{LOG}.$$

The graphical representations of conditional dependencies introduced for covariance and concentration matrices in Section 2.2 can be applied directly to binary response and intermediate response variables. There is, however, a difficulty in passing from conditional to marginal relations, as illustrated in simple form by the structure of Figure 3.4.

We can represent the model of Figure 3.4 by two regression

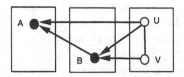

Figure 3.4 *An independence graph typical of difficulties in marginalizing in logistic regressions.*
Binary response, A; binary intermediate variable, B; two quantitative explanatory variables, U, V; $A \perp\!\!\!\perp V \,|\, (U, B)$, A directly dependent on U, B; in general no analytic expresssion for the dependence of A given just U, V available, that is after marginalizing over B.

equations. We can write these in the form

$$\Lambda\{P(B = 1 \mid U = u, V = v)\} = \alpha_{B|UV} + \beta_{BU.V} u + \beta_{BV.U} v,$$

$$\Lambda\{P(A = 1 \mid B = j, U = u, V = v)\} =$$
$$\alpha_{A|BUV} + \beta_{AB.UV} j + \beta_{AU.BV} u,$$

the term in v in the second equation being missing. The difficulty arises if one examines the marginal relation between A and U, V, i.e. marginalizing over B. If and only if we take the linear form for Λ, the limitations of which have been noted above, the marginalization is explicitly achieved in simple form and we have over a limited range, with $\Lambda(\pi) = \pi - 1/2$, that

$$\Lambda\{P(A = 1 \mid U = u, V = v)\} = \alpha_{A|UV} + \beta_{AU.V} u + \beta_{AV.U} v,$$

where, in particular,

$$\beta_{AU.V} = \beta_{AU.BV} + \beta_{AB.UV} \beta_{BU.V},$$

the formula that applies to least-squares regression coefficients, showing the regression of interest as the sum of a direct effect for fixed B and an indirect effect via the induced change in B as U varies.

These formulae are connected with general methods for representing the joint distribution of mixtures of variables of different types. In particular, the family of conditional Gaussian distributions, or CG-distributions, is formed by allowing an arbitrary distribution for a set of binary or nominal variables with, conditionally on the values of those variables, a Gaussian, i.e. normal, distribution for the continuous variables; in the homogeneous CG-distributions these Gaussian distributions all have the same covariance matrix. Then, when there is single binary variable the conditional distribution of A given the other variables is a linear logistic regression for the binary variable on the continuous ones regarded now as explanatory.

The corresponding formulae for nonlinear Λ are more complicated, although results similar to the above are available as approximations when the probabilities concerned cover a relatively narrow range.

3.7 Multivariate binary data

The above discussion deals with univariate binary responses. When the response variable is multivariate and binary, the first step, if we need a multivariate analysis, is to set out a form for the joint distribution of a sample of such variables in the absence of explanatory variables.

If we have just two binary variables A, B, say, to be treated on a symmetrical footing, it is reasonable to look for a measure of association between the components that is in a sense analogous to the correlation coefficient for multivariate normal variables. The joint distribution can then often usefully be characterized by the separate marginal distributions and the measure of association.

In many ways the most useful general measure of association is the log-odds ratio, motivated and defined as follows. Given $A = i$ the odds for $B = 1$ versus $B = 0$ are

$$P(B = 1 \mid A = i)/P(B = 0 \mid A = i) = \pi_{i1}^{AB}/\pi_{i0}^{AB},$$

say. If A and B are independent this ratio is the same at both levels of A. Hence the ratio of the separate odds ratios is a measure of association taking the value 1 for independent components. The log of the ratio is more convenient for many purposes, being zero when independence holds, so that we are led to define

$$\psi_{AB} = \log\{(\pi_{11}^{AB}\pi_{00}^{AB})/(\pi_{10}^{AB}\pi_{01}^{AB})\}.$$

Now this measure has been motivated asymmetrically as between A and B, but the explicit algebraic form shows that in fact A and B are treated symmetrically. Indeed the log-odds ratio is essentially the only measure of association with this feature. It also has the important property that its range of variation is independent of the marginal distributions of A and B provided zero probabilities are excluded.

Illustration. There has been some discussion in the political science literature as to whether the association between social class and voting behaviour has been weakening. The position in the UK appears to be that, while there have been appreciable changes in the marginal distribution of social class, the association with voting behaviour as measured by ψ has remained quite stable.   □

Suppose now that there are p component variables, $A_1, \ldots, A_p$. There are thus 2^p possible values in the joint distribution. The most general distribution is thus the saturated model which has arbitrary probabilities attached to the 2^p points, $2^p - 1$ independent parameters in all. This can be a valuable basis for exploring higher-order conditional independencies among the variables via special kinds of log-linear representations of the probabilities to be discussed below.

For $p \geq 4$ the total number of parameters in the saturated binary distribution is more than in the multivariate normal distribution, allowing more subtle dependencies to be directly explored in the binary than in the continuous case. On the other hand, in some sense there is less information in the binary than in the continuous case, leading to the suspicion that for some purposes the saturated model is overparametrized, at least for moderate or large values of p.

A special multivariate binary distribution having the same number of association parameters as the multivariate normal distribution is the quadratic exponential distribution defined as follows.

For three binary variables, conveniently denoted by A, B, C, we write

$$\log \pi_{ijk}^{ABC} = \log P(A = i, B = j, C = k)$$
$$= \mu + \alpha_A i + \alpha_B j + \alpha_C k + \alpha_{BC} jk + \alpha_{CA} ki + \alpha_{AB} ij,$$

where it leads to a more symmetrical version of the formulae to take the possible values of the random variables as -1 and $+1$ rather than the more usual 0 and 1. Here μ is a normalizing constant chosen to make the probabilities sum to one. A limitation of this representation, and of the more general form for p variables, is that if we marginalize to obtain, say, the joint distribution of A, B, or the marginal distribution of A, simple forms arise only as approximations.

Note that if a term in ijk were added we would recover the saturated model with seven independent parameters. With more components the reduction in the number of parameters is greater;

with p components the quadratic exponential model has $p(p+1)/2$ independent parameters as compared with $2^p - 1$ in the saturated log-linear model.

The conditional distribution of, say, A, B given $C = k$ is obtained by subtracting from the log joint distribution a function of k alone. The term in the product for A, B remains $\alpha_{AB}ij$ and it follows after direct substitution that the conditional log-odds ratio for A, B given $C = k$ does not depend on k and is

$$\psi_{AB|C} = 4\alpha_{AB}.$$

The general quadratic exponential takes the form

$$\log \pi_{i_1...i_p}^{A_1...A_p} = \mu + \Sigma_r \alpha_r i_r + \Sigma_{rs} \alpha_{rs} i_r i_s.$$

As before, the conditional log-odds for A_r, A_s given all remaining components is $4\alpha_{rs}$. In particular, $\alpha_{rs} = 0$ if and only if A_r is conditionally independent of A_s given all other components, and in this sense α_{rs} plays the role of the concentration in the multivariate normal case. The relation with logistic regression is that if we regard one component, say A_1, as a response and the other components as explanatory, then $2\alpha_{1r}$ is the logistic regression coefficient of A_1 on A_r when $A_2, \ldots, A_p$ are used as explanatory variables. There are thus two rather different interpretations to these *second-order parameters* α_{rs}, both involving conditioning on all other variables. The interpretation of the *first-order parameters*, α_r is less direct and usually not of particular interest, except in the absence of some interaction terms.

A difficult issue arises when we wish to include dependence on explanatory variables or more generally carry out an analysis in which the properties of components of A can be studied marginally. A linear dependence on the parameters arising directly in the quadratic exponential specification, i.e. on α_r or on α_{rs}, can be incorporated directly, for example by writing

$$\alpha_r = \beta_{r0} + x^T \beta_r.$$

This leads to a relatively simple statistical analysis but because of the indirect interpretation of α_r will usually be substantively unsatisfactory.

Typically the individual components A_r will be of intrinsic interest and the regression of their properties marginally on explanatory variables will be of direct concern. Then it will be reasonable to begin with a different formulation in which the marginal distributions have, for example, linear logistic regression on the explanatory variables,

$$\Lambda\{P(A_r = 1)\} = x^T \gamma_r$$

and, for example, to specify also the log-linear relation for the marginal association parameters. The efficient fitting of such models, called *multivariate logistic* models, is, however, more difficult.

3.8 Nominal variables

We now turn to models for the description of the distribution of variables with a small number, k, of levels, $k > 2$. We treat these levels, initially at least, as unstructured. That is, even if they are ordered or scores are attached to the different levels we start by ignoring that information.

Illustration. Blood group and eye colour, say with four or five levels, and ethnicity with, in a particular study, rather a small number of levels, are typical examples. In some contexts, a response such as symptoms recorded at three levels, absent, mild or severe, might initially be treated as nominal, although at some point the natural ordering of the levels should be acknowledged. Social class is sometimes recorded on an essentially ordinal scale, although the Goldthorpe class schema is not to be treated as ordinal.   □

Because we are dealing at this stage with nominal response variables we suppose the number of possible levels is modest; nominal variables with many levels may, however, sometimes arise as explanatory variables, especially as nonspecific explanatory variables, for example as different centres in a large study or as occupational classes in studies of occupational health.

There are several different approaches to the analysis of such responses. These are

1. to replace the variable by a nested sequence of binary variables;

2. to replace the variable by a cross-classified set of binary variables, possible directly only when k is a power of 2;

3. to attach a numerical score to each level of response chosen to capture as strongly as possible regression on explanatory variables;

4. to generalize models used in the binary case, notably the logistic model.

All these methods generalize fairly directly from the univariate to the multivariate case and so we shall concentrate here on the former.

It is hard to formalize the choice between the broad approaches listed above. If more than one leads to simple interpretable conclusions it may be necessary to report several approaches.

The first method usually requires that one level is in some sense a reference level, for example the absence of a response. Then with three levels, we might define one binary variable to indicate presence or absence of a response, and then a second binary variable, defined only for those showing a response, to indicate which of the two possible responses occurs. One part of the analysis would thus be done only on a subset of individuals, namely positive responders. This is a relatively widely applicable technique, especially when there are just three levels of response. The objective is to achieve simple descriptions of dependence on explanatory variables, possibly different for the two (or more) separate binary analyses.

In principle it might be possible with, say, three levels to do three analyses with the three possible choices of reference level and to take the analysis or analyses leading to the simplest conclusions compatible with subject-matter knowledge.

The second method is restricted to cases where the number of levels of response is a product of small integers, for example to powers of 2, in particular to responses with four levels. We then aim to represent the four levels as the result of cross-classifying two 2-level variables; the interpretation of the dependence on explanatory variables is then achieved via the constructed binary components. Again it would in principle be possible to search empirically through the three possible representations of four levels in terms of two binary components, but it seems likely that to be fruitful there should be some subject-matter basis to the construction.

We shall not in this book use method 3, namely the construction of numerical scores to attach to the levels of response.

Finally, we consider what is mathematically the most flexible and general approach, namely the formal generalization of the logistic analysis of binary responses. For some purposes, however, the interpretation is indirect. In this we write for each level, r, of response and for explanatory variable x,

$$P(A = r) \propto \exp(\beta_r^T x).$$

Because the probabilities must sum to one over all values of r, it follows that the β_r are not uniquely determined, only differences between the parameters for two different values r_1 and r_2, say, being interpretable. Alternatively, we could arbitrarily choose a reference level, say $r = 1$, and take $\beta_1 = 0$ so that the dependence on explanatory variables at other levels of response is compared with that at $r = 1$. This would be reasonable when there is one particular level of response which, on substantive grounds, can be taken as a reference level. In other applications inspection of an initial set of estimated β_r might show a simplified interpretation if a new baseline level of response were used.

One direct interpretation of the parameters is obtained by noting that, conditionally on the response being either, say, r_1 or r_2, the probability that the response is in fact r_2 is given by a binary logistic equation with parameters $\beta_2 - \beta_1$.

The distribution is closely connected with the family of homogeneous CG-distributions introduced in Section 3.7. If marginally A has an arbitrary distribution and if, conditionally on each value of A, the variables X have a multivariate normal distribution with differing means but the same covariance matrix, then, conditionally on $X = x$, the nominal variable A has the above generalized logistic form. Moreover, vanishing of regression coefficients in such a model has an interpretation in terms of conditional independence.

3.9 Ordinal variables

A form of variable common in many social science applications ranks the possible values that might be observed without, initially at least, placing particular numerical values on the different scale points. We call such variables *ordinal*; they may arise either as explanatory or as response variables. For example in a five-point scale, with scale points very bad, bad, satisfactory, good, very good, there may be no immediate sense in which the difference between bad and very bad is the same as the difference between satisfactory and bad, which would be an implication of a numerical scoring of the levels as $-2, -1, 0, 1, 2$. Hence such a five-point scale would typically be treated as ordinal.

In the use of such scales it is usually desirable as far as feasible to anchor the scale points by reference to known features or by guidance as to roughly what proportions of the responses should fall at the various scale points. The purpose is partly to avoid distributions of scores concentrated on one or two levels and thus possibly uninformative about contrasts and, more importantly, to achieve some comparability between persons.

Illustrations. In a subjective scoring system for the size of faults in a textile fabric as small, medium, large and very large it would be helpful for the observer to have in front of him or her specimens of faults to be regarded as on the borderline between the various levels of response. In a scoring of television programmes as extremely bad, bad,..., extremely good it would be helpful to give some guidance as to the approximate proportion of times the extreme categories should be used. In a self-assessed score of the frequency of pain some guidance as to the interpretation of such terms as 'very frequent' would be helpful, but the corresponding guidance for intensity of pain would be more difficult.   □

There are a number of broad ways of dealing with ordinal data. The first is to treat them as nominal. That is, in the first place the ordering is ignored, additional assurance in the conclusions resulting if the conclusions are in some sense monotonic with respect to the ordering. This may often be reasonable when there are a few levels but is increasingly unsuitable as the number of levels increases.

The second approach is to assign conventional numerical values to the scale points, typically equally spaced values such as $-2, -1,$ $0, 1, 2$ for a five-point scale, and then to proceed as for quantitative variables, typically via a linear model. Some check, at least informally, is desirable that the conclusions are reasonably expressed via contrasts of expected values and that the conclusions are not critically dependent on the scoring system used. Particular care is needed if there is some tendency to cluster around the extreme points of the scale; see Section 3.11.

Illustration. Many questionnaires used in social science investigations contain a sequence of somewhat related questions addressing a particular issue, for example health status. Often the possible answers to individual questions are ordinal, for example on a three- or five-point scale. Some questions may be so important as to merit analysis as individual responses but if there are many questions some merging is inevitable. One general strategy is to recognize on prior grounds a number of primary dimensions, for example of physical, social and psychological health status, and to accumulate a separate score within each dimension based on a simple method of scoring for each question. With a relatively small number of questions per dimension it will be feasible to check whether within a dimension there are individual questions for which the pattern of responses is different from that for the dimension total. □

When the ordinal variable is explanatory, nonlinear regression on a simple scoring system may provide a check of the suitability of the scoring scheme.

Illustration. In an investigation of patients treated in a pain clinic, ten ordinal three-point scales were used to measure different aspects of patients' experience of pain. These were used as explanatory variables in a multiple regression and nonlinearity was examined by including squared terms one at a time. For three scales nonlinearity was present and this led to some redefinition of the scoring. For example, one of the variables concerned drug usage, scored initially as

(i) -1, irregular intake of pain-relieving drugs;

(ii) 0, regular intake of at most two such drugs;

(iii) 1, regular intake of more than two such drugs.

Clear evidence of quadratic regression on this variable was found and the contributions to the fitted regression equation at levels (ii) and (iii) were virtually identical, suggesting that for this particular purpose the distinction between (ii) and (iii) was unimportant and that it would be sensible to use the contracted binary variable (i) versus (ii) and (iii) combined. □

The third approach is to develop by direct analysis scoring systems for ordinal variables that are in some sense most sensitive. The development may be on the data under current analysis or on some reference data used in setting up a measuring instrument. Thus in the first approach we might use some form of discriminant analysis to find the scoring system that expresses most sensitively the contrast of main interest. A drawback to this approach is that the scores will be different for each new set of data examined and thus comparison and consolidation of conclusions from different studies may be hindered.

Finally, we adapt the idea of conditional independence to give some simple special models for the joint distribution of ordinal variables. For simplicity suppose that there are three ordinal variables A, B, C, taking values $1, \ldots, I; 1, \ldots, J; 1, \ldots, K$.

We can form 2×2 contingency tables in two different ways. One is global. We amalgamate the data into

$$A \leq i, \ B \leq j, \ A > i, \ B \leq j; \ A \leq i, \ B > j, \ A > i, \ B > j$$

and calculate ψ_{ij}, the log-odds ratio from the four cells. A variety of models can then be formed; for example, one is that $\psi_{ij} = \psi$, a constant, for all i, j.

A different approach, more suitable when the merging of adjacent levels of response may be desired, centres on local odds ratios. We denote by $\gamma_{ii',jj'}^{AB}$ the log-odds ratio in the 2×2 subtable formed from levels i, i' of A and levels j, j' of B. Here often $i' = i+1, j' = j+1$. We omit A, B in the notation when the context is clear. This leads to consideration of the following four types of independence for two variables each at three levels:

(i) $\gamma_{12,12} = 0$, i.e. $A_i \perp\!\!\!\perp B_j$ $(i = 1, 2; j = 1, 2)$;

(ii) $\gamma_{12,12} = \gamma_{23,12} = 0$, i.e. $A \perp\!\!\!\perp B_j$ $(j = 1, 2)$, suggesting that levels 1 and 2 can be merged;

(iii) $\gamma_{12,12} = \gamma_{23,12} = \gamma_{12,23} = 0$, i.e. $A_i \perp\!\!\!\perp B$ $(i = 1, 2)$, $A \perp\!\!\!\perp B_j$ $(j = 1, 2)$ suggesting that levels 1,2 of both variables can be merged;

(iv) all four γ's vanish corresponding to $A \perp\!\!\!\perp B$.

The merging is appropriate for a concise description of remaining associations, not for studying relations with potential explanatory variables.

3.10 Mixed variables

We have discussed above a number of different types of variable mostly in their role as responses. There is no particular difficulty involved in having, for example, a nominal or binary variable as explanatory to one or more continuous variables. Sometimes, however, we need to consider as simultaneous responses variables of different types. We have already had as examples the conditional Gaussian, abbreviated as *CG-distribution*, in which there is a separate normal distribution at each level of a set of binary or nominal variables.

We now illustrate the issues involved in more detail by discussing distributions for three response variables, first taking two continuous and one binary variable and then one continuous and two binary variables. Although we are concerned with the joint distribution, it is convenient to build it up via marginal and conditional specifications. There are a considerable number of possibilities and, of course, particular applications may well suggest new forms.

First suppose that there are two continuous components, X, Y and one binary variable, A. There are a number of special independence relations of possible interest, typified by

$$X \perp\!\!\!\perp Y \mid A, \quad A \perp\!\!\!\perp X \mid Y, \quad X \perp\!\!\!\perp Y, \quad A \perp\!\!\!\perp X,$$

although it would be very unusual to be interested in all of these at the same time. The form of a suitable model for the joint distribution is determined not only by the need to be consistent with the data but also by the consideration that some independence hypotheses are much more easily examined via some models than by others and therefore a formulation must be adopted that allows the hypothesis of substantive interest to be examined. Indeed in some formulations one independency cannot hold without other apparently unrelated hypotheses being satisfied at the same time.

We shall consider just four possible models for the joint distribution.

The first form is a homogeneous CG-distribution in which the distribution of A is arbitrary and, given $A = i$, the pair (X, Y) is bivariate normal with covariance matrix Σ and with arbitrary means $(\mu_x(i), \mu_y(i))$. This is most likely to be relevant if (X, Y) are in some sense responses to A. The marginal distribution of (X, Y) is thus a mixture of bivariate normal distributions; this is, however, often hard to distinguish from a single normal distribution. The hypotheses $X \perp\!\!\!\perp Y \mid A$ and $X \perp\!\!\!\perp A \mid Y$ are easily tested and correspond to zero correlations at both levels of A and to equal regression lines in regressions of X on Y for both levels of A. Departures from independence are directly parametrized. In general, however, $X \perp\!\!\!\perp Y$ only if either $(A, X) \perp\!\!\!\perp Y$ or $(A, Y) \perp\!\!\!\perp X$, illustrating that this model would not be used if testing $X \perp\!\!\!\perp Y$ is of substantive interest, a rare case if (X, Y) are actually responses to A.

A second possibility is a homogeneous conditional Gaussian regression with A as response and with the marginal distribution of (X, Y) being bivariate normal with vector mean (μ_x, μ_y) and covariance matrix Σ. Then, given $X = x, Y = y$, the binary variable A has linear logistic regression on (x, y). In this the hypotheses $A \perp\!\!\!\perp X \mid Y$ and $X \perp\!\!\!\perp Y$ are readily tested, but not $X \perp\!\!\!\perp Y \mid A$, the hypothesis least likely to be of interest if A is actually a response to (X, Y).

Very similar to this is the model produced from a trivariate normal distribution for (X, Y, U) by dichotomizing U, thus producing a model differing from the previous one only by replacing logistic regression by probit regression.

The fourth possibility is a homogeneous conditional Gaussian regression joint-response chain model with (A, Y) as joint responses to X in which X is marginally normal and in which, given both $A = i, X = x$, Y has linear regression on X with parallel lines at the two levels of A. Thus Y is not marginally normal except in a degenerate case. Tests of $X \perp\!\!\!\perp Y \mid A$ and of $X \perp\!\!\!\perp A \mid Y$ are respectively ones of zero slopes and of identical intercepts.

Where the same hypothesis can be tested by more than one of the above models, the tests are in general not identical but are normally unlikely to differ greatly. There are also local models in which logistic or probit regression is replaced by linear regression for a limited range of values of the explanatory variables.

A rather similar discussion applies when there are two binary variables and one continuous variable. The three most commonly relevant cases are, first, a homogeneous conditional Gaussian distribution model in which the distribution of (A, B) is arbitrary and in which Y is conditionally normal with the same variance σ^2 throughout; and second, two homogeneous conditional Gaussian regressions, one with (A, B) as a response to Y and the other with A as a response to (B, Y).

In this discussion we have, for ease of exposition, put some emphasis on the testing of hypotheses of independence, but there are corresponding implications for the representation of dependencies.

3.11 A special kind of continuous data

A useful distinction when interpreting continuous data is between variables that, for given values of the explanatory variables, are clustered more or less symmetrically around a central value and those for which many individuals have values at or close to one end of the scale, that corresponding in some sense to a 'normal' state of affairs. The former, but sometimes not the latter, may be suitable for direct analysis based on normal-theory methods, i.e. means and covariance matrices may capture most of the relevant information.

Illustration. Test scores for attitude to television programmes or advertising are often based on answers to questions on five- or seven-point scales, or sums of scores over a series of related questions. The ends of the scale represent extremes of approval and disapproval that may be relatively rare, illustrating the first kind of response.

By contrast, for questions recording absence, presence and severity of adverse events of some kind, again scores, say on a five-point scale, there may be appreciable concentrations of observations at or near the 'absence' category, and covariance matrices and essentially linear regression based methods of analysis may sacrifice much information or, indeed, may mislead through failing to distinguish sufficiently sharply behaviour close to the transition level from that for more distant values. The methods of model checking set out later are designed in part to detect this possibility. □

Illustration. A further example is hourly precipitation at a particular site, which is zero with appreciable frequency. Similar remarks apply to such variables considered as explanatory.

For example, in a study of the development of infants the number of times in hospital in the first three months of life might be used as an explanatory variable; the most important aspect of this might be whether it is zero, the value when nonzero being secondary. □

If Y is a response variable with a baseline value which we take without loss of generality to be zero, it may be adequate to analyse its dependence on explanatory variables in two parts. First we define a binary variable DY to be 1 if $Y > 0$ and to be 0 if $Y = 0$ and study the dependence of $P(DY = 1)$ on x. In addition we use linear methods to relate $E_+(Y)$ to x, where $E_+(Y)$ is the conditional expectation of Y given $Y > 0$. The idea can be extended when there is a vector of responses some or all of the latter type, by using some of the ideas on variables of mixed continuous and discrete type discussed briefly in Section 3.10.

Note that in the special case where we can use a linear-in-probabilities model for DY, then

$$E(DY) = P(DY = 1) = \alpha^D + \beta^D x,$$

$$E_+(Y) = \alpha^+ + \beta^+ x,$$

and

$$E(Y) = \alpha^D \alpha^+ + (\beta^D \alpha^+ + \beta^+ \alpha^D)x + \beta^D \beta^+ x^2,$$

so that to the extent that linear forms for $E(DY)$ and $E_+(Y)$ are adequate a nonlinear form for $E(Y)$ is implied. If, however, the slopes are relatively small the quadratic term may not be detectable. If that happens and also

$$\beta^D \alpha^+ + \beta^+ \alpha^D \simeq 0,$$

then Y marginally will appear independent of x in a linear analysis and an important dependence could be overlooked. Conversely, if the dependence of $E(Y)$ on x is linear and that on the other forms nonlinear, it may be preferable to use the simpler formulation in terms of Y.

That is, in the simplest version we represent $E(DY)$ as before and write

$$E(Y) = \alpha + \beta x,$$

so that $E_+(Y)$ has the bilinear form

$$(\alpha + \beta x)/(\alpha^D + \beta^D x).$$

If a variable of the kind we are considering is used as an explanatory variable, X, say in a linear model for a continuous response Y, it may be instructive to incorporate both X and the baseline indicator DX in the linear model, writing in the simplest form for one variable, for the ith individual,

$$E(Y_i) = \alpha + \beta x_i + \gamma D x_i.$$

This can be fitted directly by standard least-squares methods. The interpretation of the parameters includes the following. If $\beta = 0, \gamma \neq 0$ then departure from baseline x changes the response but the amount of the change does not depend on the extent of the departure from baseline. If, on the other hand, $\beta \neq 0, \gamma = 0$, then the response depends linearly on the explanatory variable and the baseline is not an atypical value in this respect, i.e. a departure from baseline to $x = 1$ has the same effect as a change from $x = 1$ to $x = 2$.

Again with several such explanatory variables there are various possibilities, probably best explored in separate steps. Thus for two explanatory variables (X_1, X_2) with baseline indicators (DX_1, DX_2), the following are among the explanatory variables that might provide interpretable fits:

$$(1 - DX_1)(1 - DX_2), \quad X_1(1 - DX_2), \quad X_2(1 - DX_1).$$

If 0 corresponds not to a single value of x, but to a range of values, for example,

$$Dx = 0 \ (x < x_0), \quad Dx = 1 \ (x \geq x_0),$$

then we can represent two-phase regression by the inclusion of an interaction term, writing the model in the form

$$E(Y) = \alpha + \beta x + (\gamma - \delta x_0)Dx + \delta x Dx.$$

Here $\gamma = 0$ if the two straight lines of slopes β and $\beta + \delta$ intersect at $X = x_0$.

3.12 Bibliographic notes

Sections 3.2–3.5. Much of the material connected with the multivariate normal distribution is described in textbooks on classical multivariate analysis; see, for example, Anderson (1984) and Mardia, Kent and Bibby (1979). The interpretation of concentrations via partial correlation coefficients is due to Wermuth (1976). For a detailed treatment of factor analysis models, see Lawley and Maxwell (1971). Seemingly unrelated regression models were introduced by Zellner (1962). Models in which certain concentrations are set to zero were introduced by Dempster (1972) who called them covariance selection models. Models with some correlations set to zero are a special case of linear covariance structures analysed by Anderson (1973).

Section 3.6–3.8. Cox and Snell (1989) discuss at length the use of the logistic model for binary data and also some of the simpler extensions. Agresti (1990) gives a wide-ranging account of nominal data. For an approach to analysing sparse contingency tables via exact tests, see Kreiner (1989). Cox (1972) reviews some aspects of multivariate binary data. Glonek and McCullagh (1995) discuss the fitting of the multivariate logistic model. For the stability of class related voting in the UK, see Heath, Evans and Payne (1995). For the Goldthorpe class schema, a nominal, i.e. nonordinal, scale and special methods for its analysis, see Erikson and Goldthorpe (1993).

Section 3.9. The extensive literature on ordinal data is best approached via the book of Agresti (1984).

Section 3.11. Conditional Gaussian (CG) distributions are studied in relation to conditional independence by Lauritzen and Wermuth (1989). A range of different models for mixed binary and continuous variables were compared by Cox and Wermuth (1994b).

CHAPTER 4

Statistical analysis

4.1 Preliminaries

The previous discussion has been concerned entirely with the properties of distributions, i.e. of random variables, and not with techniques for fitting models by estimation of unknown parameters. Most of the models discussed later in this book have substantial systematic structure. Here we deal briefly with the estimation of the mean μ and covariance matrix Σ from observations on n independent individuals all with the same distribution. We assume that there are no special restrictions on the two sets of parameters.

The data are taken to form an $n \times p$ array Y. That is, the vector contributed by each individual is now written as a row. Estimates can be obtained in effect by applying to the data essentially the same averaging operations used to define the parameters. That is, we set

$$\hat{\mu}_r = \bar{Y}_{.r} = \Sigma Y_{jr}/n,$$

$$\hat{\sigma}_{rs} = \Sigma(Y_{jr} - \bar{Y}_{.r})(Y_{js} - \bar{Y}_{.s})/(n-1).$$

When deviations from a linear model, rather than deviations from a sample mean vector, are used the divisor $n-1$ is replaced by $n-d$, where d is the dimensionality of the linear model fitted.

The calculation of the sample mean and covariance matrix will rarely be more than a first step in analysis. Nevertheless, careful inspection for anomalous values or features unexpected on general grounds and the comparison of means and covariance matrices on related separate sets of data can be an important preliminary stage of interpretation.

4.2 Method of least squares

4.2.1 Formulation

In Section 3.2 we have already seen the idea of a least-squares criterion used to define a linear approximation to the conditional mean of one random variable given the values of one or more other random variables. This is concerned with the properties of random variables and probability distributions, not directly with the analysis of data.

We now turn to a more familiar use of least squares in estimation. This underlies the statistical methods of linear regression analysis and to some extent of analysis of variance. Because we assume the reader is familiar with these we give here only an outline of the key results, together with some discussion of points of interpretation.

We assume observations on a single response variable Y, all other variables being explanatory. We suppose observations available on n independent individuals and take as the underlying model that for the ith individual

$$E(Y_i) = \Sigma x_{ir}\beta_r, \ \text{var}(Y_i) = \sigma^2,$$

observations on different individuals being independent (or at least uncorrelated). Here x_{ir} for $r = 1, \ldots, q$, referring to the rth explanatory variable for the ith individual, are the observations on the explanatory variables and are treated as fixed for the purpose of the study of dependence, even though in observational studies they will usually be random. Because most models include an intercept it is commonly useful to introduce a further variable with $x_{i0} = 1$ for all i; then with $q = 1, x_{i1} = x_i$ we have the linear regression model of Section 1.6. We write d_β for the dimensionality, usually $q + 1$, of the parameter β.

There are a number of equivalent ways of writing the basic model. For example, we may write

$$Y_i = \Sigma x_{ir}\beta_r + \epsilon_i,$$

where

$$E(\epsilon_i) = 0, \ \text{var}(\epsilon_i) = \sigma^2, \ \text{cov}(\epsilon_i, \epsilon_j) = 0 \ (i \neq j).$$

In matrix notation we write

$$E(Y) = x\beta, \ Y = x\beta + \epsilon, \ \text{cov}(Y) = \text{cov}(\epsilon) = E(\epsilon\epsilon^T) = \sigma^2 I.$$

Here Y and ϵ are $n \times 1$, x is $n \times d_\beta$, and β is $d_\beta \times 1$.

We have stated a version of the error assumptions in which only variances and covariances are involved. A stronger set of assumptions needed for detailed distributional results requires the Y_i to be independently normally distributed.

4.2.2 Least-squares estimates

We define the *least-squares estimates* of β to be the values that minimize the sum, over the n individuals, of the squares of differences between observed and fitted values. A more helpful definition from a theoretical view-point is that they correspond to projecting the vector Y orthogonally onto the space spanned by the columns of the matrix x. Explicitly the estimates $\hat{\beta}$ satisfy the projection equations

$$x^T(Y - x\hat{\beta}) = 0,$$

i.e.

$$x^T x\hat{\beta} = x^T Y,$$

with solution

$$\hat{\beta} = (x^T x)^{-1} x^T Y.$$

It can be shown that, provided the original specification has no redundant parameters, i.e. provided x has full rank $d_\beta \leq n$, then $x^T x$ also is of full rank with an inverse and hence $\hat{\beta}$ is uniquely defined.

For some purposes, especially connected with the balanced models of analysis of variance, it is convenient to extend the discussion to cover models with redundant parameters, when the inverse of $x^T x$ is replaced by a generalized inverse, but we shall not do that here.

To estimate σ^2, and for some aspects of testing the adequacy of the model, we define *residuals*, $\hat{\varepsilon}$, as the difference between the observed values and the fitted values obtained by substituting least-squares estimates for unknown parameters into the model. That is, we set

$$\hat{Y} = x\hat{\beta}, \ \hat{\varepsilon} = Y - \hat{Y}.$$

Then we estimate σ^2 by the residual sum of squares divided by the *degrees of freedom* for residuals, i.e. the number of individuals minus number of parameters fitted. That is, we set

$$\mathrm{SS_{res}} = \hat{\varepsilon}^T \hat{\varepsilon}, \ \ s^2_{\mathrm{res}} = \mathrm{SS_{res}}/(n - d_\beta).$$

4.2.3 Properties of estimates

The main properties of the least-squares estimates are as follows. First, the least-squares estimates themselves are unbiased and their covariance matrix has a simple form:

$$E(\hat{\beta}) = \beta, \ \text{cov}(\hat{\beta}) = (x^T x)^{-1} \sigma^2.$$

Further, the residual mean square is an unbiased estimate of the error variance σ^2:

$$E(s_{\text{res}}^2) = \sigma^2.$$

Estimated standard errors can thus be found for any component $\hat{\beta}_r$ or any linear combination of components. Under normal-theory assumptions the Student t distribution can be used for confidence limits and significance tests. These results are approximately valid much more generally.

To test whether a particular submodel is an adequate fit we argue as follows. First, if the submodel can be defined by the vanishing of a single parameter, we examine a studentized estimate, i.e. we estimate that parameter and judge its significance by dividing by its estimated standard error, calculated as above.

If the hypothesis is in effect defined by the vanishing of some parameters γ of dimension d_γ, we fit two models by least squares, a full model with d_β parameters and a reduced model which contains d_γ fewer parameters. The difference

$$\text{SS}_{\text{res,red}} - \text{SS}_{\text{res,full}}$$

between the corresponding residual sums of squares describes the improvement in fit using the full model as compared with the reduced model. The improvement is compared with that to be expected by chance due to having d_γ additional parameters to adjust by computing the mean square for lack of fit, namely

$$(\text{SS}_{\text{res,red}} - \text{SS}_{\text{res,full}})/d_\gamma$$

and dividing it by the residual mean square from the full model to form the variance ratio, F.

If the additional parameters are in fact zero, then on average the numerator and denominator of the ratio F are equal. Statistical significance can be judged under normal-theory assumptions from standard tables of the variance-ratio distribution, i.e. the F-distribution.

4.2.4 Testing model adequacy

The general ideas involved in testing model adequacy have been sketched in Section 2.10. There are broadly three ways in which they can be applied in the present context.

The simplest and in many ways the most fruitful is, as noted previously, to add one or more terms to the original linear model representing curvature or interactive effects. These additional terms can be estimated and tested for significance.

Illustration. A model can be checked for interaction between a particular pair of explanatory variables by forming a new variable as the product of the two original variables and thereby defining a new augmented model. The simplest example, with just two explanatory variables, has the form

$$E(Y_i) = \alpha + \beta_1 x_{i1} + \beta_2 x_{i2} + \gamma x_{i1} x_{i2}.$$

The adequacy of the original model is checked by testing the hypothesis $\gamma = 0$ via the least-squares estimate $\hat{\gamma}$ compared with its standard error.

If interaction between two nominal variables, say with m_1 and m_2 levels, is to be tested it will usually be necessary to define

$$(m_1 - 1)(m_2 - 1)$$

formal product variables to represent the interaction, unless some especially meaningful components of interaction can be identified.

If dependence on a very large number of explanatory variables is analysed by a linear model, it may be impracticable to check for interaction between all possible pairs of variables. Then it may be best to examine products of only selected pairs of variables chosen partly on a priori grounds and partly by the principle that interactions of large main effects may be of most interest. Interactions between treatment variables and intrinsic variables are likely to be of particular interest. □

The second possibility is to fit an augmented model outside the set of linear models. These include models with correlated errors, with nonconstant variance, and also models in which some unknown transformation of Y obeys the linear model of interest. We shall consider the principles for analysing such models in Section 4.3.

It can be shown that any procedure for assessing the adequacy

of a given linear model is equivalent to an examination of residuals, although this may not always be transparent. A rather less
formal set of methods is based directly on the residuals which in
most respects should, if the model fitted is adequate, behave like a
set of independent and identically distributed quantities, normally
distributed when the original errors ϵ_i are normally distributed.
They can thus be plotted to check for distributional shape, plotted
against further explanatory variables and so on. The disadvantage
of this approach as contrasted with more formal approaches is that,
while a large number of checks can be generated, it is not always
easy on the basis of these to decide how to modify the initial model.

4.2.5 Weighted and generalized least squares

At a very intuitive level the method of least squares is most likely
to be effective when all the response variables Y_i have equal variance and no two response variables are strongly dependent. The
first condition justifies treating all squared errors on an equal footing when minimizing the sum of squares to form estimates and
the second condition rules out 'double counting' of errors in that
process.

These conditions are enshrined in formal optimal properties,
very strong for errors that are independent and identically normally
distributed and less strong for errors that are known only to be of
equal variance and uncorrelated.

In simple cases where the variances are unequal but in a known
ratio or, more generally, where the covariance matrix is a known
multiple of a known matrix, a fairly simple modification of the
least-squares estimates is possible.

For this suppose first that different Y_i have zero covariance and
that

$$\text{var}(Y_i) = \sigma^2/w_i,$$

where the w_i are known constants which can be regarded as weights.
We take as before the linear representation for the means

$$E(Y_i) = \Sigma x_{ir}\beta_r.$$

The simplest approach is to take as new response variables $Y_i\sqrt{w_i}$
with variance σ^2 and with new explanatory variables $x_{ir}\sqrt{w_i}$. All
the ordinary least-squares results can now be applied, leading to
so-called *weighted least-squares estimates*.

Illustration. In a simple but important example, the Y_i all have the same mean μ. This would arise, for example, when the Y_i are summary statistics estimating a treatment effect from a number of independent studies. Subject to a check of the mutual consistency of the studies, i.e. of the adequacy of the model, it may be required to produce a combined estimate of μ. The transformed model is that

$$E(Y_i\sqrt{w_i}) = \sqrt{w_i}\mu,$$

leading to the least-squares estimate

$$\Sigma w_i Y_i / \Sigma w_i.$$

That is, each value Y_i is weighted inversely as its variance.

The estimate of σ^2 derived from the residual mean square of the transformed variables is

$$\{\Sigma w_i Y_i^2 - (\Sigma w_i Y_i)^2 / \Sigma w_i\}/(n-1).$$

If the variances of the original Y_i are known or estimated with very high precision, we can compare the residual mean square directly with that known value to provide a test of the constancy of μ. □

More generally, suppose that

$$\mathrm{cov}(Y) = V\sigma^2,$$

where V is a known $n \times n$ matrix and σ^2 is an unknown constant. We can again transform to a new set of variables, uncorrelated and with constant variance σ^2, although now the procedure is slightly more complicated.

We transform to a new vector variable $Z = LY$, where the matrix L is chosen to give

$$\mathrm{cov}(Z) = LVL^T\sigma^2 = \sigma^2 I,$$

so that the components of Z have constant variance and are uncorrelated. We do not need here an explicit form for Z or for L, but if one is needed, we can proceed by successive orthogonalization to produce a lower triangular L. That is, Z_1 is taken as a multiple of Y_1, scaled to have variance σ^2. Then Z_2 is taken as a linear combination of (Y_1, Y_2) again scaled to have the required variance but also chosen to have zero covariance with Z_1. And so on.

It follows directly from the defining property that

$$V^{-1} = L^T L.$$

Now Z obeys the conditions for ordinary least-squares estimation in the linear model

$$E(Z) = Lx\beta$$

with the resulting estimate

$$\hat{\beta}_V = \{(Lx)^T Lx\}^{-1} (Lx)^T Z$$
$$= (x^T V^{-1} x)^{-1} x^T V^{-1} Y.$$

It follows by a similar argument that

$$\text{cov}(\hat{\beta}_V) = (x^T V^{-1} x)^{-1} \sigma^2.$$

Here $\hat{\beta}_V$ is called a *generalized least-squares estimate*. Weighted least-squares estimates are the special case where V is diagonal.

The immediate usefulness of these results is limited by the need to know the matrix V or the weights w_i. Sometimes, however, estimated covariance matrices or weights of reasonable precision are available and then the approximation involved in replacing, say V, by an estimate is a good one.

4.2.6 Multivariate linear model

So far in this chapter the response variable considered has been one-dimensional. Suppose that we have a p-dimensional response variable observed on n independent individuals, so that we now have an $n \times p$ matrix of random variables. In the simplest situation each component Y_r, say, follows a linear model with the same set of explanatory variables and with each component having a separate set of unknown parameters.

Thus we write

$$E(Y) = x\beta,$$

where now β is a $d_\beta \times p$ matrix of parameters, each column corresponding to a different component of response making β in fact the transpose of the matrix B introduced in the theoretical discussion in Section 3.2.2.

Because individuals are assumed independent, different rows of Y have zero covariance and we assume that each row has the same covariance matrix Σ. For stronger results we assume the different rows of Y have independent multivariate normal distributions with common covariance matrix Σ. Thus each component obeys the conditions for ordinary least-squares estimation.

It turns out that under these conditions the best estimate of β is obtained by applying the ordinary least-squares results component by component, so that

$$\hat{\beta} = (x^T x)^{-1} x^T Y.$$

A proof of this result can be obtained via an application of generalized least-squares estimation. For this we have to rearrange Y as a column vector and this is done by stacking the columns under one another to produce a $np \times 1$ vector, denoted by vec(Y). It can then be seen that

$$E\{\text{vec}(Y)\} = (I \otimes x)\text{vec}(\beta), \quad \text{cov}(Y) = \Sigma \otimes I.$$

The formulae for ordinary least squares can now be applied and after some use of the properties of the Kronecker product $\otimes$ the simple form for $\hat{\beta}$ is recovered. An alternative, more elegant, proof hinges on noting that the dimension of the unknown regression coefficients is the same as that of the right-hand sides of the least-squares equations which (together with the residual covariance matrix) form a minimal sufficient statistic.

To estimate Σ, we find the matrix of residuals,

$$\hat{\varepsilon} = Y - x\beta,$$

form the associated matrix of sums of squares and products $\hat{\varepsilon}^T \hat{\varepsilon}$ and divide by the residual degrees of freedom $n - d_\beta$.

While a series of ordinary least-squares analyses component by component can be done for other problems involving multivariate responses, there will in general be a loss of efficiency as soon as either different components have different sets of explanatory variables or constraints are imposed on the matrix of regression coefficients.

Illustration. If there are two components of response, the first regressed on, say, x_1 and the second regressed on a different variable, x_2, there will be a loss of efficiency in using two separate least-squares analyses unless the two response variables have independent residuals or the two explanatory variables are highly correlated in the data under analysis. This is the model discussed in the econometric literature *seemingly unrelated regressions* and in this book in more detail in Section 4.3.3.   □

4.3 Maximum likelihood analysis

4.3.1 Introduction

In the previous section we discussed the method of least squares, a most important basis for the analysis of empirical data. When the random part of the variation is represented by independent random variables of constant variance, and the dependence on explanatory variables is linear, the method is fully efficient, but there are many situations that are quite different, for example where the response variable is binary.

We therefore need a more general technique, preferably one that reduces to least squares in the appropriate special case. For this reason we now outline the method of maximum likelihood.

4.3.2 Description of method

We suppose that a provisional probability model is to be fitted to the data. We assume that the model specifies the distribution of the response variables in a mathematically known form but involving a number of unknown parameters denoted collectively by θ. The essential idea of the method of maximum likelihood is simple and conveyed by the name. We take the probability of the response variables evaluated at the observed values, regarded as a function of θ, henceforth called the likelihood function,

$$\text{lik}(\theta, y) = f_Y(y; \theta),$$

where $f_Y(y; \theta)$ is the joint probability or probability density of the response random variable Y.

Let $\hat{\theta}$ be the value, assumed unique, at which the likelihood is maximized. That is, for all values of $\theta' \neq \hat{\theta}$, we have that

$$\text{lik}(\hat{\theta}, y) > \text{lik}(\theta', y).$$

We call $\hat{\theta}$ the *maximum likelihood estimate* of θ.

There are many qualifications to and elaborations of this simple idea, but the general notion is direct and has intuitive appeal, at least for fairly simple forms of likelihood function. In many important practical cases to be described later in this book the likelihood can be shown to have a single maximum. If there are several local maxima the use of the method to obtain a single estimate will be unsuitable in the rather unusual case where there is not a dominant maximum; in any situation where there may be multiple maxima

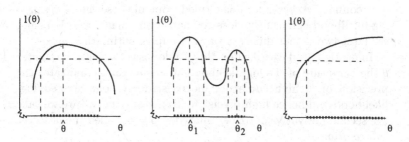

Figure 4.1 *Forms of log-likelihood function.*
(a) Single maximum; (b) multiple maxima; (c) maximum approached at infinity. Approximate confidence bands obtained from parameter values with sufficiently large log-likelihood.

care is needed to ensure that the numerical procedure finds the global maximum. Often, but not necessarily, difficulties with multiple maxima stem from badly fitting models. See Figure 4.1 for simple illustrations, showing in particular that confidence regions formed from all sufficiently large values of the log-likelihood are at least qualitatively reasonable.

When the component responses are independent the likelihood is a product. This suggests that we work with the log-likelihood, which will be a sum of contributions, and this we do from now on, denoting the log-likelihood by $l(\theta)$ or sometimes by $l(\theta, y)$. That is,

$$l(\theta, y) = \log f_Y(y; \theta).$$

Provided that the log-likelihood is a smooth function of θ at the maximum, the maximum likelihood estimate can be found by equating the derivative of the log-likelihood to zero. That is, we calculate the derivative

$$u(\theta; y) = \partial l(\theta; y)/\partial \theta,$$

often called the score function, and form the maximum likelihood estimating equation

$$u(\hat{\theta}; y) = 0.$$

As noted above, care is needed if this equation has multiple solutions.

When the amount of information about θ in the data is large, the log-likelihood will usually be quite tightly concentrated around

θ. A confidence region for θ is formed from all those values giving a log-likelihood sufficiently close to that at the maximum. We shall see later two rather different ways of implementing that idea.

In particular, if we take first the simple case of a single parameter θ the curvature of the log-likelihood at the maximum indicates the precision of $\hat{\theta}$; high curvature means strongly concentrated log-likelihood and hence high precision. To measure curvature we take the second derivative of $l(\theta)$ evaluated at $\theta = \hat{\theta}$, called the *observed information*:

$$j(\hat{\theta}) = [-\partial^2 l(\theta)/\partial\theta^2]_{\theta=\hat{\theta}}.$$

Sometimes, especially for theoretical purposes, it is more useful to work with the expected curvature, called the *expected* or Fisher *information*. This is defined by

$$i(\theta) = E\{-\partial^2 l(\theta;Y)/\partial\theta^2\}.$$

More explicitly, it can be shown that when the amount of information in the data is relatively large the distribution of $\hat{\theta}$ is approximately normal around θ with approximate variance that can be calculated as either $1/i(\theta)$ or $1/j(\hat{\theta})$. From this there results one way of calculating confidence limits (limits of error) for θ, namely as $\hat{\theta}$ plus and minus an appropriate multiple, based on the standard normal distribution, of the approximate standard error. This is often extremely convenient although a rather better method based directly on the log-likelihood function is sometimes advisable and will be described later.

More commonly, there is a vector of unknown parameters. Then the derivative of the log-likelihood, the *score function*, is replaced by a score vector of partial derivatives, i.e. by differentiating in turn with respect to the component parameters. It is convenient to use the concise notation

$$u(\theta;y) = \nabla l(\theta;y),$$

where ∇ denotes the gradient operator, the maximum likelihood estimating equation now being

$$u(\hat{\theta};y) = 0.$$

This is a set of simultaneous equations equal in number to the number, d_θ, of unknown parameters; the equations are commonly nonlinear and have to be solved by an iterative technique.

Again we usually assume either that there is a unique solution or that the overall maximum is used.

The observed and expected information are now $d_\theta \times d_\theta$ matrices with (r, s) element for the observed information

$$j_{rs} = [-\partial^2 l(\theta; y)/\partial\theta_r \partial\theta_s]_{\theta=\hat{\theta}}$$

and with a corresponding expression for the expected information matrix $i(\theta)$.

Just as in the one-dimensional case, the approximate distributional properties of the maximum likelihood estimate are determined by the information matrix. In fact, provided the errors of estimation are relatively small, we can take $\hat{\theta}$ to have an approximate multivariate normal distribution around θ with covariance matrix given by one of the inverse information matrices, $j^{-1}(\hat{\theta})$ or $i^{-1}(\hat{\theta})$. Thus, in particular, the variance of a particular component parameter is obtained by inverting one of the information matrices and taking the appropriate diagonal element. This leads as before to approximate confidence limits for the component in question.

4.3.3 Some special cases

We now give a number of relatively simple instances of maximum likelihood estimation. We put little emphasis on the elementary but often rather complicated pieces of mathematical manipulation involved.

Least-squares estimation in linear model. For the normal-theory linear model of Section 4.2 almost all of the earlier results are recovered. The log-likelihood is

$$-n \log \sigma - (Y - x\beta)^T (Y - x\beta)/(2\sigma^2),$$

where the numerator of the second term is the sum of squares of differences between observed and model values. Differentiation with respect to β, or a direct algebraic derivation, shows that the unique maximum of the likelihood occurs on minimizing the sum of squares. That is, the maximum likelihood estimate of β is the same as the least-squares estimate $\hat{\beta}$.

Direct differentiation of the log-likelihood shows, however, that the maximum likelihood estimate of σ^2 is the residual sum of squares divided by the sample size, n, not by the residual degrees of freedom $n - d_\beta$, say. This is unimportant if n is much greater than d_β, but this is not always the case, and we shall discuss later the modification to maximum likelihood required to retrieve the conventional estimate.

Evaluation of the observed and expected information matrices by differentiating the log-likelihood twice shows that for the parameter β, with σ^2 known, observed and expected information matrices are the same and the covariance matrix of $\hat{\beta}$ given in Section 4.2 is recovered exactly.

Nonlinear least squares. There are numerous models, often arising from a substantive context, in which the assumptions of the previous example hold except for the complication that the dependence of $E(Y_j)$ on explanatory variables is nonlinear. A simple example involves independent normally distributed random variables $Y_1, \ldots, Y_n$ with constant variance σ^2 and with

$$E(Y_j) = \alpha \exp(\beta x_j),$$

where the x_j are values of an explanatory variable taking at least three different values.

As before, the log-likelihood is maximized by minimizing a sum of squares, in this case

$$\Sigma\{Y_j - \alpha \exp(\beta x_j)\}^2.$$

The resulting maximum likelihood estimates satisfy

$$\hat{\alpha}\Sigma e^{2\hat{\beta}x_j} = \Sigma Y_j e^{\hat{\beta}x_j},$$
$$\hat{\alpha}\Sigma x_j e^{2\hat{\beta}x_j} = \Sigma Y_j x_j e^{\hat{\beta}x_j}.$$

These two equations are nonlinear in $\hat{\beta}$ and, atypically, linear in $\hat{\alpha}$. Iterative solution is simplified by the last feature because in effect $\hat{\alpha}$ can be eliminated. The information matrices can be evaluated in a straightforward way. The maximum likelihood estimate of the error variance is again the residual sum of squares divided by the sample size, although division by the residual degrees of freedom is to be preferred.

Multivariate linear regression. An important special problem is that of multivariate linear regression; see the previous discussion in Chapter 2. In this model, we have n independent observations on a $1 \times p$ vector Y all with the same covariance matrix Σ and such that each component of Y has linear regression on the *same* set of explanatory variables x, the regression coefficients for each component being separate independent parameters. We can write this concisely as

$$E(Y) = x\beta.$$

Here Y is now the $n \times p$ matrix of response variables, each row corresponding to an individual and each column to a component. The $n \times q$ matrix x specifies the explanatory variables, one set for each individual. Finally, the $q \times p$ matrix of regression coefficients has one column for each component variable.

If the distribution of the individual response vectors is multivariate normal the likelihood can be written down and, as noted above, the maximum with respect to β occurs at the component least-squares estimates, i.e. at the result of applying a univariate least-squares analysis component by component. The covariance matrix Σ has as its maximum likelihood estimate the matrix of residual sums of squares and products divided by the total sample size, although, as in previous examples, division by the residual degrees of freedom is to be preferred.

More generally, in an important class of multivariate models which we shall call decomposable, maximum likelihood fitting reduces to a chain of univariate least-squares analyses. This is not the case in general, however, as the next example illustrates.

Constrained multivariate regression. In the previous special case each component response had its own separate set of regression coefficients and the explanatory variables were the same for all components. In some contexts, however, we may be concerned with models in which different component responses have different explanatory variables.

If the other assumptions of the previous example remain reasonable, we can retain the specification

$$E(Y) = x\beta$$

by defining x to include *all* explanatory variables entering the specification. It then follows that certain elements of the matrix β are constrained to be zero.

While the separate least-squares analyses component by component, each with its appropriate set of explanatory variables, can still sensibly be made, there is some loss of efficiency in so doing. The maximum likelihood equations are in general different and their solution then requires iteration. A special case is examined in Section 5.3.2; for a general discussion, see Section 4.3.5.

The qualitative point involved is that, provided the components are fairly strongly correlated, the residuals from the regression of one component contain information about the errors in another component, and when the regression models are different this information can be exploited. As mentioned before, in econometrics this model is called one of seemingly unrelated regressions.

Combination of estimates. Quite generally in our discussion, the observations under analysis may be not primary data values as recorded directly on the individuals being studied but may be derived quantities, perhaps obtained from some preliminary stage of analysis. An extreme example of this arises when the same vector β of unknown parameters is estimated from two independent sets of data, perhaps by two quite different methods.

Suppose then that (Y_1, Y_2) are two independent $1 \times q$ vectors both estimating the parameter vector β with errors of estimation having approximately multivariate normal distributions with known covariance matrices (V_1, V_2). One approach is to apply the method of generalized least squares to the composite vector (Y_1, Y_2). Alternatively we can use the log-likelihood

$$-(Y_1 - \mu)V_1^{-1}(Y_1 - \mu)^T/2 - (Y_2 - \mu)V_2^{-1}(Y_2 - \mu)^T/2,$$

maximization of which yields

$$\hat{\mu} = (V_1^{-1} + V_2^{-1})^{-1}(V_1^{-1}Y_1 + V_2^{-1}Y_2),$$

with

$$\mathrm{cov}(\hat{\mu}) = (V_1^{-1} + V_2^{-1})^{-1}.$$

As with other instances of the pooling of information from more than one source, it is important to check the mutual consistency of the information being combined. In the present case, consistency is easily tested by noting that

$$(Y_1 - Y_2)(V_1 + V_2)^{-1}(Y_1 - Y_2)^T$$

has a chi-squared distribution with q degrees of freedom.

Logistic regression. The above instances centre on the normal or multivariate normal distribution, and this is in a sense atypical in that one of the main uses of maximum likelihood estimation is to deal with situations involving other forms of distribution. Consider, for example, the logistic regression of n independent binary random variables $(Y_1, \ldots, Y_n)$ on explanatory variables $(x_1, \ldots, x_q)$.

Then

$$P(Y_j = 1) = \exp(\Sigma x_{jr}\beta_r)/\{1 + \exp(\Sigma x_{jr}\beta_r)\},$$

$$P(Y_j = 0) = 1/\{1 + \exp(\Sigma x_{jr}\beta_r)\}.$$

The log-likelihood can thus be written

$$\Sigma_j y_j \Sigma_r x_{jr}\beta_r - \Sigma_j \log\{1 + \exp(\Sigma_r x_{jr}\beta_r)\}.$$

Maximum likelihood equations can now be written down on differentiating with respect to the components of β and solved by iteration.

Constrained log-linear model. A fairly common problem in the analysis of nominal data is the fitting of a log-linear model to probabilities, with some parameters in the saturated log-linear model being constrained to zero. The quadratic exponential model is one example; another arises in the fitting to ordinal data of a model containing local independencies of the kind discussed in Section 3.9.

Suppose that π_t is the cell probability for the tth cell in a contingency table, with $\theta_t = \log \pi_t$. In a sample of n independent individuals, let n_t denote the number in cell t, with $\Sigma n_t = n$; let n^T denote the row vector of cell frequencies and θ, π the corresponding column vectors of parameters.

The likelihood is obtained from the formula for the multinomial distribution of cell probabilities and the log-likelihood is then $n^T \theta$. Now log-linear representation of probabilities in the saturated model can be taken in the form

$$\gamma = C\theta, \quad \theta = D\gamma,$$

where $D = C^{-1}$ can be chosen so that the first column of D has constant elements. The log-likelihood thus has the form $n^T D\gamma$ and the first component of the parameter γ thus has constant coefficient, and the corresponding parameter could be eliminated from consideration.

Now a reduced model will have some of the components of γ set to zero, leading to a reduced matrix D_r and associated parameters γ_r. It can be shown from this that where, as is typically the case, the constraints concern vanishing interactions, the maximum likelihood estimates under the reduced model are such that the fitted frequencies in cells satisfy the no-interaction constraints and that marginal fitted frequencies agree with the observed counterpart whenever they correspond to one of the cliques in the concentration graph of the log-linear model. This is enough to determine the fitted frequencies and the corresponding estimated parameters.

4.3.4 Likelihood ratio tests and profile likelihood

The results in the previous section give a reasonably general method of point estimation parallel to the calculation of least-squares estimates discussed in Section 4.2. We now give a method for testing the adequacy of a reduced model parallel to the F test and, closely associated with this, a somewhat improved method of finding confidence limits for a parameter.

Suppose, then, that we have a full model determined by a parameter θ of dimension d_θ and a reduced model obtained after fixing the values of parameters γ of dimension d_γ. In the parallel problem for the linear model we compared the results of two least-squares fits; here we compare maximum likelihood fits of the two models having respectively maximized log-likelihoods of $\hat{l}_{\text{full}}$ and $\hat{l}_{\text{red}}$. The former is at least as great as the latter and the larger the difference the stronger the evidence of the inadequacy of the reduced model. Apparent consistency with the reduced model would, however, be illusory if the full model was itself a bad fit.

A formal test of the difference is obtained by testing

$$2(\hat{l}_{\text{full}} - \hat{l}_{\text{red}})$$

as chi-squared with d_γ degrees of freedom. The test, being based on the log of a ratio of likelihoods, is called the likelihood ratio test and the statistic itself, including the factor 2, the deviance.

A typical use of this test would be to examine in logistic regression whether a model with d_γ explanatory variables omitted gives an adequate fit. We shall see many uses of such tests, often relatively informal, in the later chapters of this book, largely to see whether a model with some special structure, for example of conditional independencies, is an adequate fit as compared with a similar form of model without the independencies.

A different use of the same distributional result is to improve on the specification of confidence limits via a maximum likelihood estimate plus and minus an appropriate multiple of an approximate standard error.

In this regard, suppose that the parameter θ defining the model is partitioned into a parameter of interest ψ and a nuisance parameter ϕ. Often ψ will be of dimension 1 or 2, corresponding to the aspect of the problem of immediate concern in one phase of the analysis. Then we define the *profile log-likelihood* $l_P(\psi)$ of ψ to be obtained by maximizing the log-likelihood over ϕ for each fixed value of ψ, considering the answer as a function of ψ. That is,

$$l_P(\psi) = l(\psi, \hat{\phi}_\psi),$$

where $\hat{\phi}_\psi$ is the maximum likelihood estimate of ϕ when ψ is held fixed; note that this is in general not the same as $\hat{\phi}$, the overall maximum likelihood estimate of ϕ.

If there are no nuisance parameters, then a profile log-likelihood function is the same as an ordinary log-likelihood.

The qualitative idea of likelihood-based confidence intervals is that all values of ψ that produce a value of $l_P(\psi)$ sufficiently close to the overall maximum are reasonably consistent with the data. More precisely, we use the approximate test stated above to 'reject' all values of ψ which are inconsistent with the data, leading to an approximate level $1 - \epsilon$ confidence interval or region defined by

$$2\{l_P(\hat{\psi}) - l_P(\psi)\} > \chi^2_{d_\psi, \epsilon},$$

where the critical value on the right-hand side is the upper ϵ point of the standard chi-squared distribution with d_ψ degrees of freedom. Thus for a one-dimensional parameter ψ, the right-hand side refers to the chi-squared distribution with one degree of freedom, derived from the square of a standard normal variable.

While this rather more complicated procedure will often give virtually the same answer as the calculation of a maximum likelihood estimate and a standard error, the method directly based on the profile likelihood is to be preferred particularly when the likelihood is of unusual shape. For example, in the unusual situation where the likelihood or profile likelihood has several nearly equal maxima, a confidence region is automatically obtained consisting of several intervals each centred on a local maximum, where a single interval may be quite misleading. See Figure 4.1b.

4.3.5 Calculation of maximum likelihood estimates

As we have seen above, the calculation of maximum likelihood estimates usually involves the solution of nonlinear equations and answers in explicit algebraic form are typically not available. For many of the most commonly used models the widely available computer packages contain iterative algorithms solving the relevant equations. In other cases a general function maximization algorithm will be used.

We shall not discuss the numerical analytic issues involved which can, however, become critical if a large number of unknown parameters are involved or if the log-likelihood surface is far from elliptical in form.

Many procedures use a Newton–Raphson procedure or some elaboration thereof.

Thus if the parameter is one-dimensional we note that if $\hat{\theta}^{(r)}$ is the rth approximation to the maximum likelihood estimate the estimating equation

$$u(\hat{\theta}) = 0$$

can be replaced by the first-order expansion

$$u(\hat{\theta}^{(r)}) + (\hat{\theta} - \hat{\theta}^{(r)})u'(\hat{\theta}^{(r)}) \doteq 0$$

suggesting the definition of an improved approximation by

$$\hat{\theta}^{(r+1)} = \hat{\theta}^{(r)} + j^{-1}(\hat{\theta}^{(r)})u(\hat{\theta}^{(r)}),$$

using the definition of the observed information as minus the differential coefficient of the score. The same definition holds for vector parameters, j then being a matrix. Convergence of this procedure yields also the observed information at the maximum likelihood point. Convergence will certainly be speeded by, and may sometimes require, a good choice of a starting value.

There are many variants of the above algorithm. The need for these arises in particular from the explicit calculation of first and second derivatives required in order to implement the Newton–Raphson procedure. In other algorithms maximization proceeds by a series of line searches evaluating the log-likelihood itself along paths designed to lead to the maximum.

The early literature on these issues put some emphasis on ease of computation, in particular on noniterative procedures. The direct need for these may seem to have been obviated by the virtually universal availability of powerful computing facilities. We see, however, some need for simpler procedures, partly to provide good starting values for iterative algorithms and, perhaps more importantly, to provide semi-qualitative insight into the connection between the data and the output of algorithms involving extensive iteration.

Such methods may range from inspection and simple calculations based on intelligent tabulation of the data and the fitting of lines by eye to various *ad hoc* numerical methods. We now describe briefly one fairly general method for obtaining a noniterative method of estimation that is very efficient provided the model fits reasonably well.

We suppose that the model to be fitted can be embedded in a larger model for which simple maximum likelihood estimates can be obtained. The larger model will often be a saturated model for covariances or multinomial probabilities for which sample covariance matrices or cell proportions are maximum likelihood estimates. As before, we shall call such a larger model the full model and the model to be fitted the reduced model. We write the parameters of the full model as $\theta = (\phi, \gamma)$, where ϕ are the parameters of the reduced model and where under the reduced model γ takes a known value, without loss of generality 0.

Suppose, then, the maximum likelihood estimate of θ under the full model is $\hat{\theta}_f = (\hat{\phi}_f, \hat{\gamma}_f)$ with information matrix i_f partitioned to correspond to (ϕ, γ). It can be shown that a close approximation to $\hat{\phi}_r$, the maximum likelihood estimate of ϕ under the reduced model, is

$$\tilde{\phi}_r = \hat{\phi}_f + i_{\phi\phi}^{-1} i_{\phi\gamma} \hat{\gamma}_f$$

with approximate covariance matrix obtained by inverting the (ϕ, ϕ) component of the information matrix in the full model. In these calculations the information matrix can be estimated under the full model and observed information used instead of expected information. Essentially $\tilde{\phi}$ adjusts $\hat{\phi}$ for linear regression on $\hat{\gamma}$ exploiting the assumption that $\gamma = 0$. The conceptual use of $\tilde{\phi}$ lies in showing how γ can be estimated from a full model by correcting in a qualitatively reasonable way.

4.3.6 Some difficulties

While the method of maximum likelihood is of wide applicability, there can be difficulties, some of which have already been mentioned and by no means all of which can be dismissed as mathematical curiosities.

The two main problems are log-likelihood functions of unusual shapes, for example with multiple maxima or approaching their supremum as a parameter tends to infinity. Here the primary difficulty is numerical; the standard iterative algorithms, especially if not used critically, may give a misleading idea of the nature of the log-likelihood function. In principle, however, once the form of the log-likelihood function is established the procedure of specifying confidence regions directly from the log-likelihood function or its profile form is in many cases a protection against major misinterpretation.

This protection is, however, not achieved in problems in which the dimension of the nuisance parameters is large when the profile likelihood of the parameter of interest is in some cases centred far from the true value of the parameter. An instance of this already encountered is the estimation of variance in a normal-theory linear model, where the maximum likelihood estimate is obtained by dividing the residual sum of squares by the sample size rather than by the residual degrees of freedom. If many regression parameters are fitted these two divisors will be quite different and the maximum likelihood estimate of variance will be a serious underestimate. Here the remedy, namely to divide by residual degrees of freedom, is simple. Rather similar issues, not so easily dealt with, arise in connection with large sparse contingency tables.

A detailed theoretical discussion of such situations is beyond the scope of this book. There are broadly two approaches. When there are many parameters of essentially the same kind, representing for example individual mean differences in data on a large number of individuals, it may be sensible to adopt an empirical Bayes approach in which these parameters are assumed to have a frequency distribution of known form with a small number of unknown parameters. This replaces the estimation of a large number of unknown parameters by that of a small number using, however, a different likelihood function.

The second approach is to separate off that part of the data that estimates the parameter of interest from that concerning the nuisance parameters and to apply likelihood calculations only to the relevant part. This is the basis of the reduced maximum likelihood (REML) method of estimating components of variance. For estimating the variance in a simple regression model it involves regarding the set of residuals as the data for analysis and leads to the conventional estimate based on residual degrees of freedom.

4.4 Testing model adequacy

We have already in Section 2.10 discussed in outline the principles involved in examining the adequacy of models. In the examples to be discussed in detail in Chapter 6 we have relied largely on the formation of test statistics for nonlinearity of regression, inserting squared terms and cross-product terms into relations such as simple linear regressions. The extra terms may sometimes form a direct basis for interpretation. More commonly, however, they serve as a warning that reformulation is necessary with some indication of the type of modification needed. Thus the presence of curvature might be evidence for the fitting of a nonlinear regression function, for transformation of one or more of the variables concerned, for removal of subsets of observations as outliers or for the special sort of representation discussed in Section 3.10.

4.5 Presentation of regression analyses

To describe the dependence of a response variable on an explanatory variable, we use some form of regression equation. The simplest, for a single explanatory variable, has

$$E(Y) = \alpha + \beta x,$$

whereas that for several explanatory variables has

$$E(Y) = \alpha + \beta_1 x_1 + \ldots + \beta_q x_q.$$

Many of the more complicated models that have been considered above, for example logistic regression equations for binary data, also have parameters that have the interpretation of regression coefficients.

We have already discussed carefully the interpretation of parameters such as β_1, specifying the increase in $E(Y)$ per unit increase in x_1 with $x_2, \ldots, x_q$ held fixed. The mathematical meaning of this is very specific and clear, although, as we have emphasized, there is much need for care in substantive interpretation.

The direct consequence of this is that in reporting the fitting of such a model, estimates of the relevant regression coefficients together with estimated standard errors provide a sensible starting point. The ratio of each to its estimated standard error, which we sometimes call the *studentized regression coefficient*, provides a t statistic for assessing statistical significance, i.e. for judging the evidence that an effect in the observed direction is indeed established by the data. Of course statistical significance is not the same as substantive importance.

There are two reasons for taking regression coefficients as a basis for interpretation. One is the operational meaning already mentioned. The other is that, at least in some contexts, the parameter is likely to be relatively independent of the distribution of the explanatory variable encountered in the particular study, i.e. of the design used. Hence regression coefficients are, by contrast with correlation coefficients for instance, likely to be relatively stable between different similar studies.

The regression coefficient of Y on X depends on the scales of measurement of the two variables and has the units of those of Y divided by those of X, for example kilograms per year if the variables are respectively mass in kilograms and time in years. This leads to readily appreciated answers when the scales involved are familiar through being in common usage. In the social sciences widely used instruments are standardized and this means that investigators in the field can appreciate the substantive importance of given changes in score. Where new or unfamiliar instruments are used it will usually be sensible to standardize the scale approximately on the data before the start of detailed analysis.

Sometimes, however, the scales used are less familiar and this makes the interpretation of a given numerical value of the regression coefficient or a given numerical change in, say, Y less immediate. Also, in any case, such questions as whether X_1 or X_2 is the more important as an explanatory variable for Y receive only an indirect answer via the relevant regression coefficient.

While these considerations do not undermine the importance of regression coefficients they do indicate an occasional need for some supplementation to ease interpretation. We achieve this by suitable standardization of the Y and X scales.

Such standardization may be external, i.e. largely determined in a way not highly specific to the data under analysis, or may be internal, i.e. determined from the data under analysis.

So far as X, an explanatory variable, is concerned, standardization is external in two main cases.

One is that the variable X is indeed measured in a way with a widely understood interpretation; examples are age in years, counts, proportions and standard test scores. Here it will be sensible to present the conclusions as a change in Y for an easily understood and relevant change, for example for a 10-year age change.

The objective should normally be to choose units such that an effect of substantive interest has a numerical value between perhaps 0.1 and 100.

The second possibility is that X is binary, or more generally that there are a set of $k-1$ explanatory variables representing contrasts between the k levels of a nominal variable. In the binary case it will usually be right to specify the relation by the difference in mean response between the two levels. In the general nominal case there may be a natural baseline level in which case the relations are best specified as the set of $k-1$ differences from baseline together with their standard errors. The difference between any two other levels is easily found; note, however, that to find the standard error of such a difference, or more generally of a more complex contrast of levels, the covariance matrix of the original differences is needed.

An alternative approach is to provide estimates of the means at all k levels of the nominal variables, setting other explanatory variables at their average levels. These means may be roughly independent so that the standard errors of the means yield at least a rough guide to the standard errors of contrasts, although a full covariance matrix is still needed for careful discussion.

When there is no suitable basis for external standardization of X it may be useful to adopt internal standardization, the simplest form of which is to take the standard deviation s_X of X in the data as the unit. For simple linear regression this involves replacing $\hat{\beta}$ by $\hat{\beta}s_X$ which has the dimensions of Y alone. We may call this and the corresponding versions in the multivariate case *semi-standardized regression coefficients*.

This modification has the advantage that the estimated change in $E(Y)$ is evaluated relative to change in X of an amount meaningful in the data under analysis and that the regression coefficients on different explanatory variables are directly comparable. Qualitatively at least, we have a direct answer to such questions as which of a number of explanatory variables has the largest effect in the data under analysis. This is not in general the same variable as the one with most extreme statistical significance. A major disadvantage is that the conclusions from different sets of related data are not in general comparable unless the dispersions of explanatory variables nearly agree in different studies.

There is the further possibility of internal standardization of the response variables producing dimensionless numbers to describe the dependence. In some ways this is most suitably done by taking as the unit of Y the standard deviation residual to a well-fitting model. It is in many ways easier, however, and is common practice, to standardize by the marginal standard deviation of Y. The resulting rescaled regression coefficient is often called the *standardized regression coefficient*.

For simple regression this replaces the regression coefficient of Y on X by $\hat{\beta}s_x/s_Y$, the correlation coefficient. In general we obtain a multiple of the relevant partial correlation coefficient, not necessarily between limits -1 and 1.

Such internal standardization can be valuable for appreciating the consequences of fitting a particular regression model to data and, especially, for comparing the fits of different models to the same data. They are in general not suitable for comparing the conclusions from different sets of data. Because one of the primary purposes in reporting analyses is to allow future investigators to use the conclusions and to make comparisons between studies, it follows that reporting standardized regression coefficients on their own will rarely be satisfactory.

Illustration. In Section 6.2 a main dependent variable is Y, glycosylated haemoglobin, a measure of glucose control in diabetic patients. This varies from 5 to 14, with values up to about 7 or 8 corresponding to reasonable metabolic adjustment. This is a widely used scale in that field and further standardization of it is not needed. One explanatory variable used is duration, W, of illness in years and again no further standardization is called for. A regression coefficient of Y on W is directly interpretable. Another explanatory variable is X, a score for knowledge about the illness. A significant regression coefficient of Y on X of -0.061 is obtained; does this correspond to a substantively important effect? The instrument used to obtain X is less familiar and here there is a case for standardizing X, i.e. multiplying the regression coefficient by the marginal standard deviation of X in the data to obtain the semi-standardized value of -0.44 with a standard error of 0.22. Thus over a range of plus or minus two standard deviations, roughly the span of the data, a change in Y of just under 2 is involved and, especially near the critical level of 7 or 8, such a difference could be of substantive importance. Thus the qualitative conclusion is that, if confirmed at its estimated value, the regression on X could be of substantive importance but only if a full range of variation of X is contemplated.   □

Some additional considerations arise when the explanatory variable is a derived variable, such as the square of a primary variable included to capture nonlinear effects or the product of two, or more, primary variables included to represent interactions. The best way to present the former is typically to choose three reference levels for X, which for internal standardization could be the data mean of X and the mean plus and minus one data standard deviation provided that X is roughly symmetrically distributed. Provision of three means is one simple way by which the qualitative interpretation of curvature, and of its amount relative to the slope present, can be judged. The response variable may or may not be internally or externally standardized.

Estimates representing interactions are rarely best considered on their own; the composite effects of the explanatory variables involved and the interaction term must be considered. For two quantitative variables X_1 and X_2 and their product $X_1 X_2$ representing linear by linear two-factor interaction, a simple procedure is to choose two response levels of each variable and, for the four resulting combinations of the two variables, to find the fitted mean response, taking any other explanatory variables at their mean levels. The resulting 2×2 table of means may then suggest a more specific interpretation. If internal standardization is used the four reference levels would be $(\bar{X}_1 \pm s_{X_1}, \bar{X}_2 \pm s_{X_2})$, where $\bar{X}_j$ and s_{X_j} are the mean and standard deviation of X_j in the data, although for highly unbalanced data this can be unsatisfactory because one or more of the reference levels may be remote from the data. Similar considerations apply to higher-order interactions.

When the response is binary, or more generally nominal, analysis will often be via a linear logistic model. Then the regression coefficients defining the dependencies are interpreted as the change of logistic transform of the probability π of a positive response, i.e. the change in $\log\{\pi/(1 - \pi)\}$ or log-odds, per unit increase in the explanatory variable X. With familiarity, changes in the log-odds can be interpreted directly, especially when the probability involved is small, or near one, in which case the difference of logistic transforms is close to the log of a ratio of probabilities. In other cases, however, it is helpful, especially for multiple regression, to set all explanatory variables except one equal to standardized values, for example the data mean or baseline value, and then to set, say, X_1 equal to two externally or internally standardized values and to specify the estimated dependency directly in terms of the corresponding estimated probabilities or to plot its effect keeping other explanatory variables at typical values as will be illustrated in Section 6.2.

An issue related to but distinct from the interpretation of regression coefficients is the measurement of overall adequacy of fit. Where internal standardization of a quantitative response variable is unnecessary, fit may be best assessed directly by the magnitude of the residual standard deviation. Moreover, this will often be the most reasonable basis for comparing the fits achieved with different sets of data. Where internal standardization is desirable, comparison of the residual standard deviation of response to the marginal standard deviation can be used. Except for corrections for degrees of freedom in estimating standard deviations, this ratio is essentially equivalent to one minus the squared multiple correlation coefficient, R^2. This last is defined as the proportion of the variation of Y accounted for by the model.

For simple regression with one explanatory variable this reduces to use of the squared correlation coefficient, which has the stated interpretation. In multiple regression, because each partial correlation coefficient squared is a proportional reduction in variance remaining, it follows that

$$1 - R^2 = (1 - r_{Y1}^2)(1 - r_{Y2.1}^2)(1 - r_{Y3.21}^2)\cdots,$$

where, for example, $r_{Y2.1}$ is the partial correlation coefficient between Y and X_2 given X_1. This shows that as a new explanatory variable is added, the largest proportional reduction of $1 - R^2$ is achieved when the new variable has maximum partial correlation with response given the variables already in the model.

An interpretation of R^2 must, however, be made cautiously. If the objective is prediction in the narrow sense, that is prediction of the responses of new individuals with a range of values of the explanatory variables roughly similar to that in the data, then comparison of variances of prediction is of direct appeal. Otherwise, quite apart from the dependence on internal standardization, what appear to be modest values of R^2 can correspond to differences that are substantial for many purposes. Thus, if linear regression is used for binary responses, or alternatively the definition of R^2 is generalized to cover logistic regression, it can be shown that quite small values of R^2 around 0.1 are the largest that are likely to occur. Or again, suppose that the response is continuous and that there is a single binary explanatory variable taking its two values with equal probability. Then a difference of $2k$ standard deviations between the two group means induces

$$R^2 = k^2/(k^2 + 1),$$

so that, for example, $k = 1/2$ gives $R^2 = 0.2$, apparently a small number. Yet for some purposes such a difference in mean response would be of major importance. On the other hand, if one of the explanatory variables is a previous measure of the response, then the multiple correlation coefficient must become large merely because of stable measurements.

4.6 Bibliographic notes

Section 4.2. Most textbooks on statistical theory and methods, especially those emphasizing regression, have a discussion of the method of least squares. The mathematical level can vary greatly. Weisberg (1980) is a good introduction.

Section 4.3. Again most intermediate text-books on statistical theory give an introduction to maximum likelihood theory such as Hogg and Craig (1978). Cox and Hinkley (1974, Chapter 9) give a moderately thorough account and Barndorff-Nielsen and Cox (1994; see especially Chapters 3 and 4) emphasize recent developments. For REML see Patterson and Thompson (1971).

Section 4.4. Recent work on testing model adequacy has tended to emphasize so-called regression diagnostics based on residuals; see Atkinson (1985) and Cook and Weisberg (1982). The approach here is close to Cox and Wermuth (1994c). For a connection between cumulants in the joint distributions and tests in regressions of square and interactive terms, see Cox and Wermuth (1993b).

Section 4.5. For a discussion of the multiple correlation coefficient for binary data, see Cox and Wermuth (1992b).

CHAPTER 5

Special methods for joint responses

5.1 Preliminaries

In this chapter we review the analysis of some special models that are important both in their own right and as illustrative examples. The emphasis is on situations where more than a sequence of univariate analyses is needed. As we have mentioned previously, decomposable independence structures are expressible via a system of univariate regressions and can be analysed via a sequence of univariate analyses, often least-squares regressions or logistic regressions. The graphical representation is equivalent to a directed acyclic graph. When the most efficient analysis is not equivalent to such a series of univariate analyses it is, of course, desirable that the extra complications of any more elaborate analysis undertaken should lead to a genuine gain in interpretation.

The detailed methods to be used depend on the type of the response variable, as well as on any special information about the plausible form of the regression relation. Response variables can for instance be continuous, particularly those to which normal-theory regression methods are reasonably applicable, binary, nominal and ordinal. We shall concentrate largely, although not exclusively, on the first two.

There are two broad situations in which no very special multivariate considerations arise. One, as has been repeatedly stressed above, is where the independence hypotheses are decomposable. The other is that of essentially saturated models.

The advantage of univariate recursive regression systems lies not only in simplicity of analysis but also, in some ways even more importantly, in that each zero parameter, i.e each edge missing in the graphical representation, corresponds to a clearly defined conditional independency between two variables and each edge present to a conditional dependency of substantive interest between one

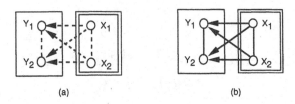

(a)   (b)

Figure 5.1 *Graphical representation of two types of saturated linear regression models for two responses Y_1, Y_2 and two explanatory variables. (a) General linear model, each response having the same set of explanatory variables, X_1, X_2. (b) Block regression, Y_1 regressed on Y_2, X_1, X_2, and Y_2 regressed on Y_1, X_1, X_2. Double lines indicate distribution of explanatory variables not specified by the model.*

variable regarded as a response and another variable regarded as explanatory, all taken conditionally on the other explanatory variables. Further, the recursive representation is at least suggestive of a process that could have generated the data. In this sense *multivariate regression graphs* (dashed-line joint-response chains of two boxes) also have some appeal in that it is not difficult to set up processes that involve hidden variables and that could have generated the residual dependencies involved. This is different for *block-regression graphs* (full-line joint-response chains of two boxes).

5.2 Joint-response regression: some saturated models

We start with the simplest instance of a saturated model for the regression analysis of continuous, approximately normally distributed random variables. This is the multivariate general linear model, already outlined in Section 4.2.6. The representation by a complete graph is shown in Figure 5.1a for two response variables and two explanatory variables; the ideas extend immediately to the general case of d_y response variables Y and d_x explanatory variables X. The variables X are contained in a double-lined box to indicate that their values are regarded as fixed. Provided the regression is linear, the full matrix of regression coefficients unconstrained and the covariance matrix of errors Σ constant, separate least-squares analyses, component by component, are efficient, as explained in Section 4.2.6.

Another set of separate least-squares analyses applies if one node or set of nodes of X is in effect disconnected from the response variables, i.e. if the corresponding regression coefficients for all components of response are zero. The corresponding null hypothesis can be tested by a likelihood ratio test, involving a ratio of determinants in the test statistic

$$\frac{n}{2} \log(\det \hat{\Sigma}_{\text{red}} / \det \hat{\Sigma}_{\text{full}}),$$

where the two matrices $\hat{\Sigma}$ are respectively those of residual covariance matrices from the reduced and the full model. The reduced model has a particular explanatory variable or group of explanatory variables independent of all response variables. The full model will often be the saturated model. Twice this statistic can be tested approximately as chi-squared.

The simplest situation for checking whether the covariance matrix is constant is where the individuals can be arranged in groups, individuals in a group having essentially the same value of all relevant explanatory variables. Then we can estimate a separate covariance matrix within each group. If the covariance matrices change systematically with the explanatory variables a subject-matter interpretation may be possible and in any case the precision of between-group comparisons may be affected.

At times the residual dependence between component response variables is of direct interest and then more than univariate analyses will be required. This can be for two rather different reasons. Major changes in the residual covariance matrix will affect the efficiency of estimates of regression coefficients for the same reason that in the univariate case changes in variance affect the efficiency of ordinary least-squares analyses. The second and more interesting possibility is that changes in the covariance or concentration matrix are of intrinsic interest, i.e. can be given a substantive interpretation.

Methods for assessing changes in marginal variances in univariate problems are widely available.

Illustration. If s^2 is an estimate of variance based on d degrees of freedom, then $\log s^2$ is approximately normally distributed with variance $2/d$ and this forms the basis of formal tests and plotting procedures. The method is rather inefficient if d is less than about 5. Nonnormality of the underlying distribution changes the variance by a factor of $(1 + \gamma_2/2)$, where γ_2 is the standardized fourth cumulant of the underlying distribution.   □

In the light of whatever changes in variance may be detected, it is necessary, especially if the number of component variables is large, to decide what particular features of the dependence should be studied. Examples are correlation coefficients, covariance elements, certain residual regression coefficients and, as measures of overall variability, the *generalized variance*, i.e. the determinant of the covariance matrix. It is therefore useful to have some of the simpler properties of such statistics as a basis for plotting and more formal procedures.

Illustration. If r is an estimated correlation based on d degrees of freedom, then

$$z = \tfrac{1}{2} \log\{(1 + r)/(1 - r)\}$$

has approximately a normal distribution of mean

$$\zeta = \tfrac{1}{2} \log\{(1 + \rho)/(1 - \rho)\}$$

and variance $(d - 2)^{-1}$, where ρ is the population correlation. The correlations between the different correlation coefficients in a situation with three or more variables can be found but have quite a complicated form and will not be used here.

The same results apply also to partial correlations obtained from a sample concentration matrix; the degrees of freedom are reduced by one for each response component eliminated. □

Illustration. If interest is concentrated on the correlation between a particular pair of variables, direct use is possible of the z-transform defined above. For example, the value of z in two sections of data can be compared and the difference tested for significance. More generally, if it is suspected that the correlation may depend on some further feature of the data, for example a third variable, one procedure is to divide the data into groups, each group having approximately the same value of that third variable, and to compute the values of z group by group. This is now in effect estimating the partial correlation given the variable that defines the groups. If the resulting values of z differ significantly, a particular form of dependence has been detected. In general, if the grouping is defined by one or more explanatory variables, z can be regressed on these explanatory variables. Comparison of the residual mean square of z with its theoretical value, $(d - 2)^{-1}$, examines whether all the variation in the correlation can be accounted for by the regression relation set up for z. □

Illustration. If, for a p-dimensional variable, $\hat{V}$ is the generalized variance based on d degrees of freedom, and V is the corresponding population value, then approximately

$$E(\log \hat{V}) = \log V$$

and

$$\text{var}(\log \hat{V}) = 2 \sum_{j=1}^{p} (d - j)^{-1}.$$

Thus with a number of groups of observations it may be suspected that the general level of variability is greater in some groups than in others. $\square$

If there is a general tendency for variances to change with mean levels, transformation component by component, for example by taking logs, may be helpful. If such transformations also remove nonlinearity, they will also increase and possibly stabilize correlation coefficients.

In many ways an easier approach, applicable whenever variances or covariances are likely to vary systematically with the levels of one or more explanatory variables, is to adapt the simple nonlinearity tests outlined in Section 4.4. For instance, if for particular variables (X, Y) the regression of X on Y^2, adjusting for regression on Y, is large and it is sensible to take Y as a response to X, then there is evidence that the conditional variance of Y changes systematically with X. Similarly appreciable regression of U on XY, adjusting for linear regression on X and Y, where (X, Y) are regarded as responses to U, is evidence that the covariance of X and Y changes systematically with U. When this is found empirically some special form of detailed model will be needed; see, for example, Section 3.11.

The corresponding discussion when all the component response variables are binary is complicated by the nonavailability of specifications that retain simple form and interpretability over marginalization. To take the simplest case, suppose that there are just two component response variables (A_1, A_2) and that each has linear logistic regression on the same set of explanatory variables x, so that

$$\log\{P(A_1 = 1)/P(A_1 = 0)\} = \alpha_1 + \beta_1^T x,$$
$$\log\{P(A_2 = 1)/P(A_2 = 0)\} = \alpha_2 + \beta_2^T x,$$

the structure corresponding to the linear regression for the component means in the continuous case. To complete the specification, for example of constant odds ratios, the simplest possibility arises when the data are arranged in groups with constant values of the explanatory variables within each group and if each group has a separate value of the odds ratio of A_1 and A_2. Then separate fitting of marginal logistic models is appropriate.

On the other hand, if we assume that the odds ratio for (A_1, A_2) is a constant, ψ, say, for each fixed x, then the individual probabilities $P(A_1 = i, A_2 = j)$ do not have a simple form so that direct application of the method of maximum likelihood is not straightforward. Further, it is not clear how much loss of efficiency is involved in estimating the parameters β from the separate logistic regressions.

Maximum likelihood estimation is, however, needed if fully efficient analyses are required comparing the fit of the above model with that of other candidate models.

These results generalize to deal with situations with more than two explanatory variables and to mixed responses.

5.3 Two nondecomposable independence hypotheses

5.3.1 Preliminaries

We now discuss two models that contain nondecomposable independence hypotheses and for which therefore fully efficient analysis involves more than standard univariate analyses, even under the simplest normal-theory assumptions.

The first is that of seemingly unrelated regressions. Figure 5.2a (see also Figures 3.1 and 3.2) shows the simplest case of two response variables and two explanatory variables; the structure of the chordless 4-chain, indicating the independencies $Y \perp\!\!\!\perp U \,|\, V$ and $X \perp\!\!\!\perp V \,|\, U$, amounts in the simplest case of normal-theory multivariate linear regression to constraining the matrix of regression coefficients by requiring two zero elements in the appropriate places.

That model is to be contrasted with the chordless 4-cycle in concentrations summarized in Figure 5.2b. Here the conditional independencies are $Y \perp\!\!\!\perp U \,|\, (V, X)$ and $X \perp\!\!\!\perp V \,|\, (U, Y)$ corresponding to vanishing concentrations in the equivalent concentration graph.

The first model could arise as an initial hypothesis about how the data might have been generated. This is not so likely for the second model. See, however, Section 2.5 for an interpretation in

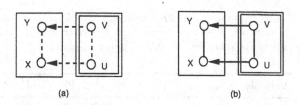

(a)   (b)

Figure 5.2 *Two different linear regression models for response variables* (Y, X) *with explanatory variables* (U, V), *both not general linear models. (a) Seemingly unrelated regression model with* $Y \perp\!\!\!\perp U \,|\, V$ *and* $X \perp\!\!\!\perp V \,|\, U$. *(b) Block regression model with* $Y \perp\!\!\!\perp U \,|\, (V, X)$ *and* $X \perp\!\!\!\perp V \,|\, (U, Y)$ *being independence-equivalent to concentration graph with the same skeleton, that is to a full-line chordless 4-cycle.*

terms of latent response variables on which conditioning occurs. Both models can arise also as reduced models, that is as simplifications suggested by initial analysis of the data.

We examine the models in turn both for continuous responses and for binary responses.

5.3.2 Seemingly unrelated regressions

We consider the simplest case with pairs of response variables (X_i, Y_i) independently bivariate normally distributed with means $(\beta u_i, \gamma v_i)$ and covariance matrix Σ. The discussion is changed only very slightly if unknown intercepts are included in the regression relations.

The log-likelihood is

$$\frac{n}{2} \log(\det\Sigma^{-1}) - \sum (X_i - \beta u_i)^2 \sigma^{11}/2$$

$$- \sum (X_i - \beta u_i)(Y_i - \gamma v_i)\sigma^{12} - \sum (Y_i - \gamma v_i)^2 \sigma^{22}/2,$$

where the σ^{ij} are elements of the inverse covariance matrix.

Differentiation with respect to β, γ gives the maximum likelihood equations. With Σ known, these are the generalized least-squares equations.

Similarly, the maximum likelihood estimate of Σ is then the matrix of residual sums of squares and products from the fitted model divided by n. Thus the full maximum likelihood solution when all parameters are unknown can be obtained by iteration

Table 5.1 *Correlations among four variables measured before an operation for 44 patients. Marginal correlations (lower triangle), partial correlations given the other two variables (upper triangle).*

Variable	Y	X	V	U
Y, log syst/diast bp	1	-0.57	-0.24	0.30
X, log diast bp	-0.54	1	-0.11	0.49
V, body mass	-0.25	0.34	1	0.57
U, age	-0.13	0.51	0.61	1

starting with the ordinary least-squares estimate for β, γ and their residual sum of squares and products, to produce revised estimates of the regression coefficients and so on. There is no guarantee that the procedure converges.

If the residual correlation is small the gain in efficiency is small. In fact, the variance of an estimated regression coefficient is in the full analysis reduced from that in separate least-squares regressions by the factor

$$(1 - \rho^2)/(1 - \rho^2 \rho_{uv}^2).$$

Here ρ is the residual correlation between the response variables (X, Y) and ρ_{uv} is the correlation between the explanatory variables (U, V).

Illustration. The correlations and partial correlations shown in Table 5.1 show no simple structure involving zero correlations or concentrations, but are in fact consistent with the independencies $Y \perp\!\!\!\perp U \,|\, V$ and $X \perp\!\!\!\perp V \,|\, U$, which, V and U being explanatory variables, is of the seemingly unrelated regression form.

5.3.3 Covariance selection model

In some respects the fitting of the analogous covariance selection model based on zero concentrations is easier. Here, with four variables we have $Y \perp\!\!\!\perp V \,|\, (X, U)$ and $X \perp\!\!\!\perp U \,|\, (Y, V)$. The technical reason for the relative simplicity is that the model corresponds to a hypothesis about parameters occurring directly in the log-likelihood of the data, parameters called in the theoretical literature canonical parameters of an exponential family distribution. It is a consequence that the following procedure for estimating the covariance matrix converges to the unique solution of the maximum likelihood equations.

(i) Estimate covariances corresponding to elements unconstrained in the concentration matrix by the standard sample covariances.

(ii) Find, by iteration, values for the remaining covariances which are such as to force to zero the required elements of the concentration matrix.

If in (ii) we cycle through the entries to be estimated in turn, each individual equation is a linear equation which can thus be solved via two evaluations plus an interpolation. In the special case of just one concentration zero only a linear equation has to be solved for the single unknown, but with two concentrations zero quintic equations are involved and in general no closed form solution is available. For numerical solution in practice, however, we proceed differently, using cyclic fitting or algorithms similar to iterative proportional fitting.

5.3.4 Binary variables

The previous discussion has been based on covariance or concentration matrices, i.e. has assumed essentially linear relations and is particularly suited for normally distributed data. We now consider more briefly the parallel discussion for binary variables.

For this we consider four variables A, B, C, D having the quadratic exponential distribution

$$\log \pi_{ijkl}^{ABCD} = \mu + \alpha_A i + \alpha_B j + \alpha_C k + \alpha_D l$$

$$+\alpha_{AB} ij + \alpha_{AC} ik + \alpha_{AD} il + \alpha_{BC} jk + \alpha_{BD} jl + \alpha_{CD} kl,$$

where for symmetry i, j, k, l take values 1 and -1. On marginalizing over A, i.e. summing over the two values of i, we have that

$$\log \pi_{jkl}^{BCD} = \mu' + (\alpha_B j + \alpha_C k + \alpha_D l + \alpha_{BC} jk + \alpha_{BD} jl + \alpha_{CD} kl)$$

$$+ \log \cosh(\alpha_A + \alpha_{AB} j + \alpha_{AC} k + \alpha_{AD} l).$$

If we now expand the last term by Taylor's theorem, ignoring cubic terms, we recover a quadratic exponential model, where, for example, the new value of α_{BC} is

$$\alpha_{BC(A)} = \alpha_{BC} + \alpha_{AB} \alpha_{AC} \text{sech}^2 \alpha_A.$$

Thus the conditions $A \perp\!\!\!\perp D \,|\, C$ and $B \perp\!\!\!\perp C \,|\, D$ for the seemingly

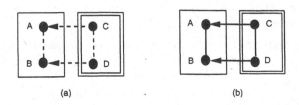

(a)   (b)

Figure 5.3 *Two different regression models for binary response variables* (A, B) *with binary explanatory variables* (C, D).
(a) Analogue to seemingly unrelated regressions with $A \perp\!\!\!\perp D \,|\, C$ *and* $B \perp\!\!\!\perp C \,|\, D$. *(b) Block regression with* $A \perp\!\!\!\perp D \,|\, (B, C)$ *and* $B \perp\!\!\!\perp C \,|\, (A, D)$ *being independence-equivalent to concentration graph with the same skeleton, that is to a log-linear model represented by a full-line chordless 4-cycle.*

unrelated regression model are represented approximately by two nonlinear relations for the original parameters of the joint distribution of A, B, C, D, namely

$$\alpha_{AD} + \alpha_{AB}\alpha_{BD}\mathrm{sech}^2\alpha_B = 0, \quad \alpha_{BC} + \alpha_{AB}\alpha_{AC}\mathrm{sech}^2\alpha_A = 0.$$

For the analogue of the covariance selection model the discussion is much more direct. For the 'mixed' terms in the quadratic exponential model give the log-odds ratio in the conditional distribution of one pair of variables given all the remaining variables. Thus the conditional independencies, say $A \perp\!\!\!\perp C \,|\, (B, D)$ and $B \perp\!\!\!\perp D \,|\, (A, C)$, are given directly by the vanishing of the parameters α_{AC}, α_{BD}.

Illustration. In a study of 350 children chosen so that at birth there were approximately equal numbers of children with organic and psycho-social risk factors, the variables A, B were psychic disorder (yes, no) and motoric handicap (yes, no) at four years, C, D being corresponding variables at two years. The variables C, D are marginally associated; if, however, psychic disorder at four years is to be predicted from the other three variables, D, motoric development at two years, does not improve the prediction, i.e. $A \perp\!\!\!\perp D \,|\, (B, C)$ and similarly $B \perp\!\!\!\perp C \,|\, (A, D)$. See Figure 5.3, where two different models, analogous respectively to seemingly unrelated regression models and to block regression models, are shown. They have different interpretations but the same independency structure shown by a full-line chordless 4-cycle.   □

5.4 Analysis of joint-response chain models

Joint-response chain models provide a quite flexible family of representations of dependencies for systems in which the variables can, from subject-matter considerations, be arranged in more than two blocks as responses, intermediate responses and purely explanatory. There are a number of associated statistical problems, namely to fit a model of specified structure, to show the departures from such a model so that the nature of discrepancies can be assessed and to compare the fits of competing explanations.

The simplest procedure for fitting a specified type of model is to fit each component response variable in turn, finding an appropriate set of explanatory variables for each, including possible nonlinear terms. We then examine the residual dependence structure of the component responses, for example by estimating the covariance and concentration matrices of the residuals from the separate regressions when the component responses are all quantitative. Note that residuals from block regressions estimate minus the corresponding concentrations. If both response variables are binary, the analogous procedure is to estimate the conditional log-odds ratio for the two variables after eliminating regression on the explanatory variables.

By such procedures estimation can be achieved via standard univariate procedures, at least approximately. Except in special circumstances efficient analysis will require maximum likelihood fitting from first principles, typically requiring special algorithms. Again, approximate solutions are readily found for models being closely supported by the data; the approach sketched in Section 4.3.5 via augmentation to a full model provides one route.

Illustrations of these ideas will be given in Chapter 6.

5.5 Bibliographic notes

Section 5.2. The multivariate general linear model handled by a separate univariate least-squares regression for each component response is described in detail in all the major books on multivariate analysis; see, for example, Mardia, Kent and Bibby (1979). For special issues relating to conditional independence hypotheses, see Cox and Wermuth (1993a).

Section 5.3. Seemingly unrelated regressions were introduced by

Zellner (1962) and have been fairly widely used in econometrics; block regression was introduced by Wermuth (1992) for reinterpretation of Gaussian full-edge chain models. Gaussian concentration-graph models, that is full-line graph models, are due to Dempster (1972) and have been called covariance selection models.

A method of fitting reduced models based on augmentation to a saturated model was discussed by Cox and Wermuth (1990). There is a general issue concerning the loss of efficiency in using separate least-squares when the efficient analysis is more complicated. Zellner and Huang (1962) discussed this for seemingly unrelated regressions.

Section 5.4. The quadratic exponential binary model was suggested by Cox (1972) and in the context of spatial processes on a lattice by Besag (1974) and was studied in more detail by Cox and Wermuth (1994a). Glonek and McCullagh (1995) studied multivariate logistic models. For applications to longitudinal data, see Zhao and Prentice (1990), and Fitzmaurice, Laird and Rotnitzky (1993).

CHAPTER 6

Some specific applications

6.1 Preliminaries

In the earlier chapters we have from time to time given outline illustrative examples. These were included to exemplify very specific technical points and the background details of the examples were therefore kept to a bare minimum. We now describe a number of applications in greater depth. While it is not feasible to explain all the considerations that enter into the analysis and interpretation of relatively complex sets of data, not least because of the crucial significance of background subject-matter considerations, we have tried in this chapter to give enough detail both to make the examples intrinsically interesting and to form the basis for some general conclusions about strategies for analysis set out in Chapter 7.

6.2 Glucose control

Complications arising for patients with insulin-dependent diabetes can be largely avoided if the patients receive instead of a conventional therapy a form of intensified treatment, in which the appropriate dose the patient needs is made to depend on the level of blood glucose as measured several times per day. When the effectiveness of several different types of intensified treatment was explored it became apparent that factors other than medical may play an important role for the compliance of a patient with the suggested treatment. Therefore a pilot study with patients receiving conventional treatment was designed at the University of Mainz to identify psychological and socio-economic variables possibly important for glucose control.

From this study we give in the Appendix data for 68 patients with fewer than 25 years of diabetes. The variables considered are shown in Figure 6.1. Glucose control, Y, is measured by glycosylated haemoglobin (GHb). A score for knowledge about the illness, X, and three different attitudes of the patient to the illness are all

SOME SPECIFIC APPLICATIONS

136

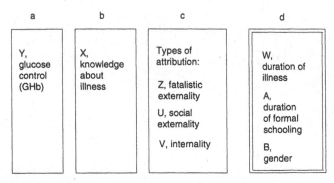

Figure 6.1 *First ordering of variables for diabetic patients.*
Variable Y, glucose control, is response variable of primary interest; X, knowledge about illness, is response of secondary interest and intermediate variable: potentially explanatory for glucose control and also potential response to psychological aspects (box c) and to background variables (box d). Variables measuring patients' type of attribution Z, U, V are treated on equal footing; background variables W, A, B are intrinsic patient characteristics, purely explanatory variables and treated on equal footing; double-lined box indicates relations taken as given.

measured with questionnaire scores. The attitudes are called fatalistic externality, Z (mere chance determines what occurs); social externality, U (powerful others are responsible); and internality, V (the patient him or herself is responsible for how the illness develops). The available background information about the patient is duration of illness in years, W; level of education, A (1, at least 13 years of formal schooling, 44%); and gender, B (1, females, 51%).

The observed range for glycosylated haemoglobin is from 5 to 14, with values up to about 7 or 8 indicating good glucose control, i.e. reasonable metabolic adjustment. Among the 68 patients are some whose illness had just been diagnosed and others who had been ill for many years. They all have had at least 10 years of formal schooling; 44% obtained the '*Abitur*' (German certificate when leaving high school), so that they have had at least 13 years of formal schooling.

Means, standard deviations, minimum and maximum values observed are shown in Table 6.1, and correlations are given as a summary of some aspects of the bivariate distributions. Table 6.1 has the general form recommended in Section 3.4 in which marginal

Table 6.1 *Summaries of marginal and of bivariate distributions for 68 diabetic patients. Marginal correlations (lower triangle), partial correlations given all remaining variables (upper triangle).*

Variable	Y	X	Z	U	V	W	A	B
Y, gl. con.	1	−.24	.09	−.13	.09	−.25	−.29	−.05
X, knowl.	−.34	1	−.34	−.08	.04	.00	.17	.14
Z, fat. ext.	.15	−.49	1	.41	−.26	.28	−.00	.09
U, soc. ext.	.03	−.32	.52	1	−.09	−.08	−.12	−.14
V, intern.	.04	.14	−.33	−.23	1	.16	−.07	−.23
W, dur. ill.	−.12	−.11	.28	.10	.05	1	−.25	.04
A, school.	−.32	.33	−.26	−.20	−.01	−.25	1	−.14
B, gender	−.07	.09	.08	−.06	−.22	.07	−.09	1
min.	5.4	11.0	8.0	8.0	27.0	0	−1	−1
max.	14.1	46.0	33.0	42.0	48.0	24	1	1
mean	9.3	35.4	19.0	24.2	41.3	10.4	-	-
st. dev.	2.0	7.3	5.4	7.0	4.7	7.0	-	-

correlations are shown below the diagonal and partial correlations given the six remaining variables above the diagonal. For a study of all variables on an equal footing the large entries in the upper half would correspond to edges which are present in the concentration graph of all variables.

For the particular ordering of the variables given in Figure 6.1 only the partial correlations displayed in the first row are relevant for interpretation. There are three moderately large partial correlations with Y. They indicate that glucose control is predicted to be better, that is GHb values are lower, the higher the knowledge about illness, X, the longer the duration of illness, W, and if patients had longer formal schooling, A at level 1. Each statement holds for patients comparable with respect to the remaining six variables.

The procedure followed to study the relations among the variables via separate regression analyses is summarized in the regression graphs of Figure 6.2. For instance, for X as response, the six potential explanatory variables are reduced to two. For the two responses of main interest, Y and X, none of the variables U, V, B remains as an important explanatory variable in the regression

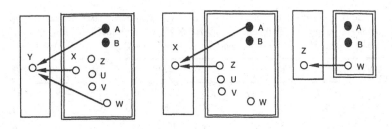

Figure 6.2 *Results of three separate regression analyses represented with the help of regression graphs.*

equation. In describing regression equations we use the notation common for generalized linear models. That is, in writing $A * W$ we imply that the regression of Y contains main effects A, W and an interaction $W.A$. Whenever an interaction term is included so too must be the corresponding main effects.

The regression equation of Y was in outline established as follows. A linear regression with main effects showed small effects, of at most one standard error, for the explanatory variables Z, U, V, B, in line with the partial correlations of row 1 of Table 6.1. Addition of squared terms, one at a time, to this regression gave no evidence of important contributions. Examination of the cross-product terms showed one large contribution with $t = 3.56$ for the interaction $A.W$. This was included in the selected regression model as summarized in Table 6.2. Note that there is a quasi-linear dependence of Y on W and A. Even though there is an interactive effect, there are also nonvanishing main effects. These are reflected in the nonzero partial correlations for (Y, W) and (Y, A) in Table 6.1, computed by ignoring the interactive effect.

Note also that the regression coefficients in the reduced model are almost unchanged from their values in the full model; their formal standard errors are only slightly reduced so that the conclusions about the major explanatory variables are essentially the same in the simplified model. This stability would, for instance, not be expected if there were major near-collinearities in the explanatory variables.

The value of the squared multiple correlation coefficient is $R^2 = 0.34$, reasonably large by the standards usual in this field of study, especially since no variables capturing the general health and med-

Table 6.2 *Regression equations for Y , glucose control.*

Response				Explanatory variable				
Y	X	Z	U	V	W	A	B	$W.A$
(a) on all explanatory variables and one interaction term								
est. coeff.	−.06	.02	−.02	.05	−.04	−.53	−.08	.11
st. error	−.03	.05	.04	.05	.03	.23	.22	.03
ratio, t	−1.89	.35	−.71	.98	−1.12	−2.33	−.35	3.48
*(b) on X and A * W*								
est. coeff.	−.061	-	-	-	−.034	−.524	-	.111
st. error	.030	-	-	-	.031	.221	-	.031
ratio, t	−2.02	-	-	-	−1.10	−2.37	-	3.56

$W.A$ is the interaction represented by the product $(W - \bar{W}) \times A$;
$A * W$ consists of main effect terms A, W and the interaction $W.A$

Table 6.3 *Selected regression equations for X, knowledge about illness, and for Z, fatalistic externality.*

Response	Expl. variable			Response	Expl. variable	
X	Z	A	const.	Z	W	const.
est. coeff.	−.578	1.610	46.592	est. coeff.	.214	16.805
st. error	.144	.786	-	st. error	.092	-
ratio, t	−4.01	2.05	-	ratio, t	2.33	-

ical status of the patients are included in the set of explanatory variables.

Table 6.3 summarizes more briefly the regression equations derived for the dependence of X and Z on their explanatory variables. We have now reduced the description of the data to three simple regression equations. It remains to interpret what has been found.

First, to describe the dependence of Y on X, A and W, we note that the effect of X is weak in the sense that a change in knowledge of two data standard deviations is estimated to achieve an improvement of about one unit in Y.

For the dependence on A and W where an interaction is involved, we look separately at the regressions of Y on W at the two levels of A; see Figure 6.3. The interpretation is that the relation is negative in one case and positive in the other, the correspond-

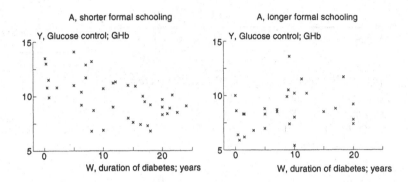

Figure 6.3 *Conditional dependence of glucose control, Y, on duration of illness, W, given level of formal schooling, A.*

ing correlation coefficients being −0.56 and 0.30. More specifically, there is a substantial difference in the initial phases of the illness with those having less schooling having the higher scores, that is a poorer metabolic adjustment.

The detailed description is completed by interpreting in a similar way the regression coefficients of X on Z and A and of Z on W as in Table 6.3.

The scales of both X and Z are somewhat arbitrarily fixed by the chosen questionnaires and in this case supplementation by the fully standardized coefficient can be helpful. Considering just patients with a comparable duration of illness, W, the standardized regression coefficient of −0.43 means that knowledge, X, is predicted to be about one standard deviation lower for patients with a score in fatalistic externality, Z, about two standard deviations higher.

Finally, we have regarded W and A as purely explanatory and hence not considered their joint distribution as a primary object of study. Nevertheless, we examine briefly their association. Because A, formal schooling, is binary, we may examine the mean durations of illness, W, at the two levels, 11.9 yr and 8.4 yr. The difference between these is quite appreciable as such and compared with the marginal standard deviation of W, namely 7.0 yr. The latter comparison of mean differences is essentially equivalent to use of the correlation coefficient of W and A, which is here −0.25, but the difference is more directly interpretable.

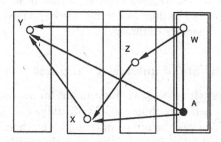

Figure 6.4 *Independence graph for just those variables directly related to*
Y, X, the two responses of main interest.
Individual regressions summarized in notation for generalized linear
models as $Y : X + A * W, \quad X : Z + A, \quad Z : W.$

Figure 6.4 provides a univariate recursive regression graph as
a summary of the sequence of the regression analyses. For each
response it shows the directly explanatory variables and implies
conditional independence from the remaining potential explanatory variables.

The representation does not, however, show explicitly the interaction term involved. As described above, the latter is reflected
in the notation for generalized linear models, the direction and
strength of effects being obtained from the estimated regression
coefficients. In particular, a qualitative summary interpretation of
the relations is as follows.

Glucose control is better the more a patient knows about the
illness and, for patients with a shorter formal education, the longer
the illness lasted. Knowledge is better the longer the formal education and the less a patient is convinced that mere chance determines what occurs. Fatalistic externality in turn is higher with a
longer duration of illness. A special feature of the patients under
study is a longer duration of illness for those with shorter formal
education.

In summary, a relatively simple interpretation of the data has
been achieved by fairly intensive analysis. Especially because of
the quite small number of patients, independent confirmation of
the conclusions is desirable. Further studies, both in Mainz and in
Düsseldorf, have confirmed the importance of social and attitudinal factors in achieving good metabolic adjustment. In particular,

programmes have led to good glucose control in which possible effects of therapy and patient behaviour are explained in detail soon
after the first diagnosis of illness.

6.3 Determinants of university drop-out

To develop a psycho-diagnostic instrument for counselling prospective students when they choose a field of study, 3500 German pupils
were selected one year before completing high school between 1973
and 1975. About 73% (2544) of the students enrolled in university
degree programmes and these formed the group for investigation.

Responses were recorded to a number of psychological questionnaires and tests and to questions regarding school career and
demographic background. It was also recorded whether students
successfully completed their studies or dropped out of university.
The data collection ended in 1984 with 2375 students still in the
study. We used data for 2339 students having complete records on
ten variables.

The main aim is to identify important developments in the
school and student career which might increase the risk that a
student stops studying without having received a university degree.

Figure 6.5 shows a first ordering of the variables under study
in a chain of boxes which reflects knowledge from previous studies
and judgement in this context about responses, intermediate variables and purely explanatory variables, based largely on the time
sequence involved. The dependence chain specified by Figure 6.5
implies that we study the relations between the variables in five
conditional distributions, that of variables in box a on those in all
other boxes, of variables in box b on those in boxes c, d, e, f and
so on. The variables in box f we treat as purely explanatory with
associations not to be specified by the model.

There are several binary variables: A, university drop-out; B,
change of high school; C, poor integration into the high school
class; D, a high school class repeated; E, change of primary school;
F, education of the father. In each case coding is 1 for yes and
-1 for no. Three questionnaire scores are available as quantitative
information, all measured during a student's first year at university: Y, self-expected achievement, the student's expectation of his
achievement in the field; X, motivation, the student's motivation
towards high achievement in the field; Z, integration, the student's

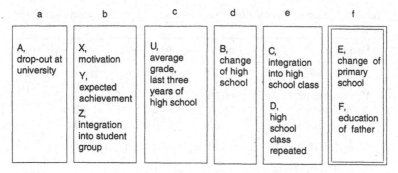

Figure 6.5 *First ordering of variables for university students.*
Variable A, drop-out at university (box a), is response variable of pri-
mary interest; for instance, B, change of high school (box d), is an inter-
mediate variable, potentially explanatory for dropping out at university
(box a), for the student's attitudes towards his study situation, (box b)
and for grades at high school (box c) and also a potential response to
other school career and demographic variables (boxes e, f). Variables in
boxes b, e are treated on equal footing, because no direction of dependence
is specified; in box f because they are purely explanatory.

perceived integration into his student group at university. The variable U is a measure of a student's performance in high school, the
mean of final marks at high school averaged over all subjects and
years 11–13. In each subject excellent performance is rewarded with
a '1' and the worst mark is '6'. Table 6.4 summarizes the marginal
distributions of the variables under study.

To decide whether main effect regressions are likely to give a
good description of the dependencies we examine normal probability plots for cross-product terms in linear regressions and for
large three-factor interactions in all marginal 2^3 tables formed after median-dichotomizing the quantitative variables. This last was
done because of the substantial number of binary variables in the
study. The plots of Figure 6.6 did not lead to evidence of interactions likely to affect interpretation. All larger t values in Figure
6.6a, estimated in this screening phase for binary responses from
linear-in-probability regressions, turned out to correspond to much
smaller t values in the appropriate logistic regressions.

Figure 6.6b shows that not a single three-factor interaction is
large in a marginal 2^3 table, all t statistics being well within the

Table 6.4 *Summaries of marginal distributions for 2339 university students.*

Variable	mean	st. dev.	min.	max.	% yes
A, drop-out	-	-	-	-	14.6
Y, achievement	6.14	2.09	0	8	-
X, motivation	35.27	8.72	10	60	-
Z, integration	6.45	2.44	0	9	-
U, av. mark, yrs 11–13	2.81	0.54	1.1	4.4	-
B, high sch. change	-	-	-	-	20.9
C, poor sch. intgr.	-	-	-	-	10.3
D, sch. class repeat	-	-	-	-	34.2
E, prim. sch. change	-	-	-	-	20.1
F, father *Abitur*	-	-	-	-	42.7

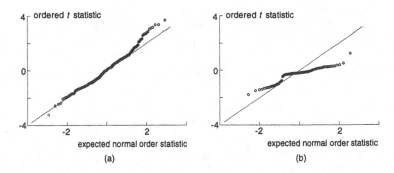

Figure 6.6 *Normal probability plots of t statistics for variables on university drop-out.*
(a) For cross-product terms, t from trivariate regressions such as regressing Y on X, V and $X \times V$. (b) For three-factor interactions, t from marginal $2 \times 2 \times 2$ tables after median-dichotomizing quantitative variables.

range of -2 to 2. The deviation from the line of unit slope just reflects high correlations of results in some of these tables.

For a general summary of the pairwise relations between variables, somewhat analogous to the full correlation matrix for quantitative variables, we give in Table 6.5 studentized regression coef-

Table 6.5 Checks of bivariate relations for variables related to university drop out. Studentized regression coefficients in linear and logistic regressions. Marginal relations (lower triangle), partial relations given all remaining variables (upper triangle). Responses are the variables on the left.

Variable	A	Y	X	Z	U	B	C	D	E	F
A, drop-out	—	-6.03	-.74	-5.51	6.56	.39	-.62	.57	.64	-1.21
Y, achievement	-9.23	—	9.03	11.41	-4.27	-.67	-.59	.43	.32	-.04
X, motivation	-4.91	12.12	—	7.29	-11.18	-.93	-.88	.97	.64	-1.63
Z, integration	-8.20	13.26	8.05	—	-2.19	-4.01	-3.47	-.79	-2.53	-.85
U, av. mark, 11-13	8.53	-7.29	-12.00	-3.17	—	.53	2.52	17.17	-.29	-3.32
B, high sch. change	2.14	-2.50	-2.49	-5.00	3.91	—	1.12	10.10	3.90	4.01
C, poor sch. intgr.	.94	-2.23	-2.37	-4.01	3.78	2.29	—	4.33	.68	1.61
D, sch. class repeat	3.97	-2.63	-3.67	-2.96	18.00	10.28	4.34	—	2.06	-.95
E, prim. sch. change	1.07	-.52	-.15	-3.20	.07	4.88	1.10	1.94	—	6.48
F, father Abitur	1.26	-.32	-1.07	-1.52	-3.29	4.36	1.66	-.68	6.49	—

ficients in linear and logistic regressions, that is t statistics. Note that these give statistical significance, not directly strengths of relation.

There is a larger number of statistics in the lower half of the table which are greater than 2 in absolute value. Thus, there are more significant dependencies in the marginal bivariate distributions than for variable pairs considered partially given all remaining variables. This corresponds to an overall covariance graph having more edges present than the concentration graph and suggests that there are several indirectly or common explanatory variables for some of the responses.

The form of the conditional distributions is investigated with separate regression analyses simplified by the procedure illustrated in the previous example. In particular, for the primary response A, university drop-out, a logistic main effect regression with three of the eight potentially explanatory variables is chosen.

Table 6.6 shows the resulting regression equations. The included explanatory variables all have highly significant effects and it has been checked that in each case there is no further explanatory variable or quadratic or interactive term whose inclusion would improve the fit considerably. The relations form a quasi-linear system since only main effect regressions remain. Note that in studying the variables X, Y, Z a block regression approach has been used in which for each component the other two components serve as explanatory variables.

We interpret these equations in two steps. The first, essentially qualitative, is via independence graphs. It would be possible with this number of variables to give a single graph, but it appears clearer to present separate graphs for the university variables given the high school variables and for the high school variables on their own.

These are shown in Figures 6.7 and 6.8, respectively. A short qualitative interpretation of the graphs is that drop-out is directly influenced by the attitudinal variables Y, Z and by academic performance in the final years of high school as measured by the average mark, U, and that these variables in turn depend on some of the earlier properties.

The second step is to interpret the regression equations of Table 6.6 more quantitatively, in particular the logistic regression for A. One way of appreciating the sizes of the individual terms in the equation is via Figure 6.9 which shows the estimated probability

Table 6.6 *Selected regression equations for variables related to drop-out.*

Response		Explanatory variables					
A	Y	Z	U	const.			
est. reg. coeff.	−.169	−.136	.962	−2.774			
st. error	.027	.024	.130				
ratio, t	−6.3	−5.7	7.4				
Y	X	Z	U	const.			
est. reg. coeff.	.044	.195	−.350	4.313			
st. error	.005	.017	.078				
ratio, t	9.1	11.6	−4.5				
X	Y	Z	U	const.			
est. reg. coeff.	.775	.363	−3.364	37.634			
st. error	.085	.072	.321				
ratio, t	9.1	5.0	−10.5				
Z	X	Y	B	C	E	const.	
est. reg. coeff.	.028	.272	−.232	−.250	−.164	3.351	
st. error	.006	.024	.060	.079	.060		
ratio, t	4.9	11.4	−3.9	−3.2	−2.7		
U	C	D	F	const.			
est. reg. coeff.	.043	.194	−.035	2.903			
st. error	.017	.011	.010				
ratio, t	2.5	17.7	−3.4				
B	D	E	F	const.			
est. reg. coeff.	.540	.239	.215	−1.09			
st. error	.053	.061	.053				
ratio, t	10.3	3.9	4.0				
C	D	const.					
est. reg. coeff.	.298	−2.10					
st. error	.069						
ratio, t	4.3						
D	C	E	const.				
est. reg. coeff.	.295	.099	−.366				
st. error	.07	.054					
ratio, t	4.3	1.9					
E	F	const.					
est. reg. coeff.	.338	−1.362					
st. error	.052						
ratio, t	6.5						

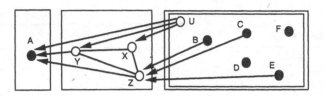

Figure 6.7 *Independence graph for university variables given high school variables and demographic background variables.*
All selected regressions are main effect regressions, written in the no-tation for generalized linear models, e.g. as $A : Y + Z + U$, *and* $Y : X + Z + U$.

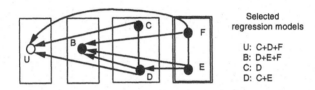

Selected
regression models

U: C+D+F
B: D+E+F
C: D
D: C+E

Figure 6.8 *Independence graph for high school variables given demo-graphic background variables.*

versus two of the three explanatory variables, in each plot holding the other two variables fixed at their means.

The estimated probability for not obtaining a university degree decreases from about 0.30 for students expecting themselves to achieve poorly in the chosen field to about 0.10 for students expecting the highest achievement; it increases from about 0.02 for students with the best average marks in the last years of high school to about 0.40 for worst average marks. Both statements are for students with about average values in the other two directly explanatory variables.

The partial dependence plot of A on Z, which is not shown, looks very similar to that of A on Y since the partial regression coefficients are nearly identical and an about equal range of values has been observed for Y and Z. We have in Section 2.7 emphasized the care needed in giving regression coefficients a substantive interpretation, and of course these plots are subject to this caution.

As noted above, in studying the variables in box b of Figure 6.5, namely Y, X, Z, we have used block regression and found contributions to each component from the other two. This analysis

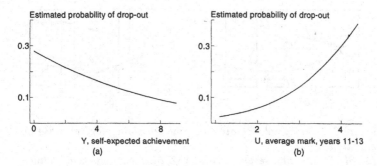

Figure 6.9 *Fitted probability of university drop out, A.*
Response A versus (a) achievement, Y; (b) average high school mark,
U. In both plots other variables fixed at mean; estimated equation is in
Table 6.6.

preserves the three variables on an equal footing as does a multi-variate regression approach. Since in the block regressions Y and X are both directly related to Z as well as to U an additional dependence of Z on U is present in the multivariate regressions, that is after marginalizing over the other joint responses, Y and X.

It is relatively speculative to introduce special relations between the components but, partly because X is not directly explanatory for the primary response A and partly because of the simplification achieved, we have explored treating Y as a response to Z and to U, ignoring X, that is marginalizing over it. Figure 6.10 shows the resulting structure which in terms of implied remaining independencies is unchanged from the original one.

The average mark over the last three high school years, U, is better, that is lower, with no high school class repeated, with good integration into the high school class and if the father had had the same type of education, only the effect of class repetition being sizeable.

The predicted rate for changing high school, B, is highest if the student has repeated a high school class, if the father had had formal schooling of 13 years or more, and if the student had already changed primary school. The corresponding estimated probabilities of changing high school differ appreciably, ranging from 0.44 to 0.11.

For variables C, D, E, F all associations are in the expected direction: a pupil is more likely to change primary school at least

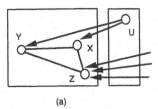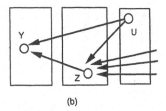

(a)   (b)

Figure 6.10 *Marginalizing over a variable exploited for simplified inter-pretation.*
(a) Block regression results with Y, X, Z treated on an equal footing. (b)
Compatible univariate recursive regressions after marginalizing over X,
with Y response only to Z, U and Z response to U and three further
variables.

once if the father has longer formal schooling and a student is more likely to repeat a high school class if he does not integrate well into this class and if he had already changed primary school. However, none of these relations is very strong; the last is included because it proved to be significant in other studies.

In addition to the separate regression results an important aspect of the chain-graph representation is that it shows a number of indirect paths to the response of primary interest, permitting additional interpretations. For instance, the path from X to A via Y is consistent with the following interpretation: motivation for high achievement, X, is likely to increase the confidence in high achievement in the field, Y, which in turn reduces the risk of dropping out from university, A. The path from E to D to B to Z to A can be summarized as follows: change of primary school, E, increases the risk that a high school class will have to be repeated, D, which in turn increases the risk that the student will change high school at least once during his school career, B. Once a high school change has been experienced it becomes less likely that a student integrates well into his later student group, Z, and this in turn is a direct risk factor for A, leaving university without having obtained a degree.

The possibility of using the graph to infer additional independencies and induced dependencies by the methods of Section 8.5 is an additional appealing feature which will be developed there.

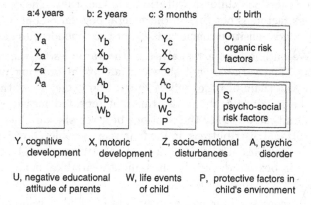

Figure 6.11 *Ordering of variables in prospective study of child development.*
Order corresponds to times of measurement (boxes a to c). Four aspects of child development with Y, X, Z quantitative, A binary. Stacking of boxes indicates variables independent given all variables in boxes to the right and double lines indicate variables held fixed. Here marginal independence of organic and psycho-social risk at birth fixed by design.

6.4 Disturbances in child development

To investigate the genesis of motoric, cognitive, socio-emotional and psychic disturbances in young children a prospective study was designed at the Zentralinstitut für Seelische Gesundheit in Mannheim. One hundred newborn children were first included in the study without attention to special risk factors, then a further 262 children were selected so that roughly the same number (about 40 children) were observed in each of nine possible risk classifications (low, medium, high) of two sum-scored variables, organic risk, O, and psycho-social risk, S.

Among the factors thought to be organic risks are low gestational age, low birth weight, asphyxia during birth, and gestosis, an illness in pregnancy. Among the factors considered to be psychosocial risks are psychic disorders of a parent, constant disharmony in the marriage and poor housing. The same group of psychologists and physicians observed and recorded a wide range of different aspects of child development at ages 3 months, 2 years, 4 years. In 1995 the first results for children aged 8 years will become available and almost 350 children will still be in the study. The distribution of risk factors is not that in the general population but is intended to provide sensitive conclusions about the dependencies of interest.

We outline here a reanalysis of the variables for the first four years. The variables are shown in an initial ordering in Figure 6.11 as a chain of boxes reflecting our knowledge about the design of the study and about the times of repeated measurement. It implies that we study the relations between the variables in three conditional distributions, in the four-year variables, box a, given all other variables in boxes b, c, d; the two-year variables, box b, given those in boxes c, d; and the three-month variables, box c, given the baseline variables in box d.

We consider four different aspects of child development, cognitive, Y; motoric, X; socio-emotional, Z; and psychic, A. The first three are recorded as sum scores. The last is binary giving a physician's yes–no judgement on whether the child shows signs of psychic disorder. As possible explanatory variables for the child development there are three sum scores based on questionnaires, interviews with parents and standardized observation of behaviour. They are called negative educational attitude, U, of the parents; life events, W, that is experiences of heavy strain for the child; and protective factors, P, against developing disorders, such as a close mother–child relation, good psychic health of parents and socially secure environment for the child. Each quantitative measurement is reported as a rescored value. That is, it is standardized with respect to the mean and standard deviation of the 100 children who entered the study first, without selection regarding risk factors.

Since one of the purposes of the analysis is to isolate early important predictors of disturbances in child development, we do not include variables as potential regressors for a response if they are recorded at the same time as the response in question. In particular, we first investigate the four different aspects of child development (Y, X, Z, A) separately after having checked for outliers and nonlinearities.

The screening for nonlinear relations among the quantitative variables gives strong evidence that main effect regressions are unlikely to give a good description of all dependencies. Figure 6.12a is a probability plot of the test statistics for the inclusion of squared terms in the regressions and shows extremely strong evidence of nonlinearities. These have to be interpreted, which we do by look-

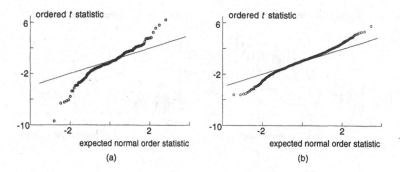

Figure 6.12 *Normal probability plot of t statistics for variables on child development.*
(a) Plot for squared terms, bivariate regressions such as Y on X and X²; some very large absolute values point at nonlinear relations. (b) Plot for cross-product terms, t from trivariate regressions such as regressing Y on X, V and X × V.

ing at scatter plots such as Figure 6.13, and a route then chosen to take account of the nonlinearities in further analysis.

Figure 6.13a in fact shows the relation between cognitive development of the child at age 4 years, Y_a, and cognitive development at age 2 years, Y_b. The nonlinearity is accounted for by severe disturbances, that is by those children having low standardized values. If children with such a handicap are looked at separately, that is if cases with standardized values below -1.5 are set aside, the relation between the two developmental variables among the remaining cases is nearly linear. To achieve this split in the regression analyses we define further binary variables: cognitive handicap (1, yes;-1, no) at ages 4 years, DY_a; 2 years, DY_b; and 3 months, DY_c. For the same reason and in a similar way motoric handicap for the three measurements is defined to give binary variables DX_a, DX_b, DX_c.

A different form of nonlinearity arises in relating the number of days a child is hospitalized up to 3 months to the number of organic risk factors present at birth as shown in Figure 6.13b. The nonlinearity is largely related to the number of children with no hospitalization. In particular, almost none of the children without any organic risk factor was hospitalized. This explains the strong nonlinear regression of organic risk on the number of days hospitalized detected in the screening plot. The number of days of

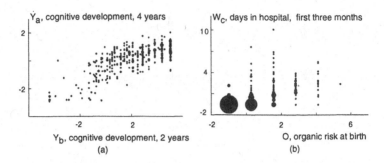

Figure 6.13 *Scatterplots, illustrating different forms of nonlinearity.*
(a) Plot for cognitive development at four years, Y_a, versus correspond-
ing variable at two years, Y_b; relation for children with very poor de-
velopment (values less than -1.5) different from relation for remaining
children. (b) Plot for days in hospital during first 3 months, W_c, versus
organic risk at birth, O; nonlinearity due to different variability. Area
of bubbles proportional to number of cases.

hospitalization is regarded as a measuring life events at this young
age. We do not model this change in variation directly but instead we define an additional binary variable, hospitalized up to 3
months (DW_c, 1 for yes, -1 for no) to describe two qualitatively
quite different situations for child development; then, only if there
is hospitalization does the number of hospitalizations measure the
severity of the strain to the child.

The normal probability plot for the inclusion of cross-product
terms in Figure 6.12b points, in addition, to some large interactions
present in marginal trivariate distributions of the quantitative measurements. Thus, mere main effect regressions are unlikely to give
best-fitting equations.

Table 6.7 checks whether there are sizeable linear components
for the four responses at 2 and at 4 years in bivariate relations.
The signs of the associations are all in the expected directions
for the chosen coding of variable categories. Measurements at 4
years of cognitive and motoric development, Y_a, X_a, and of socio-
emotional development, Z_a, are strongly related to those at 2 years.
This indicates that they are fairly stable measurements at that age.
Psychic disorder, A, is weakly related to previous measurement no
matter between which time points it is considered, suggesting less
reliable measurements.

Table 6.7 *Linear components in bivariate relations for selected variables releated to child development. Studentized regression coefficients in linear or logistic main effect regressions.*

Expl. var.	Response variable							
	Y_a	Y_b	X_a	X_b	Z_a	Z_b	A_a	A_b
Y_b	19.9	-	12.3	-	−6.6	-	−4.9	-
X_b	9.6	-	16.1	-	−2.6	-	−2.2	-
Z_b	−5.0	-	−4.1	-	10.9	-	7.1	-
A_b	−4.7	-	−3.8	-	8.1	-	6.0	-
U_b	−5.2	-	−2.8	-	8.5	-	6.4	-
W_b	−2.4	-	−3.0	-	3.9	-	3.4	-
DY_b	−14.4	-	−10.0	-	4.8	-	3.8	-
DX_b	−8.9	-	−13.0	-	3.5	-	2.7	-
Y_c	6.3	7.2	5.7	6.0	−3.8	−3.5	−3.4	−2.9
X_c	3.9	4.5	6.5	7.8	−2.1	−3.6	−2.1	−3.1
Z_c	−3.6	−4.9	−4.4	−4.1	3.8	3.5	3.0	2.6
A_c	−3.3	−4.2	−4.3	−3.6	3.8	2.1	3.5	1.2
U_c	−5.3	−6.1	−3.1	−3.2	5.9	5.9	5.0	3.7
W_c	−5.3	−5.9	−6.2	−7.1	2.7	2.3	2.3	1.9
P	5.6	6.2	4.0	4.6	−7.2	−5.5	−5.2	−4.5
DY_c	−6.2	−6.2	−5.6	−5.8	3.2	3.0	2.3	2.6
DX_c	−4.6	−4.4	−5.4	−5.4	2.0	0.7	2.4	1.2
DW_c	−3.9	−4.5	−4.2	−4.7	1.9	2.7	1.7	3.6
O	−3.8	−3.9	−5.8	−7.1	2.3	2.7	1.5	2.9
S	−5.6	−6.4	−4.2	−3.5	6.2	4.4	4.5	3.4

We now follow the procedure for model development essentially as for the previous examples. Because of the complexity of the study we give only some outline results. Table 6.8 gives the selected regression equations for cognitive and motoric development at 4 and 2 years.

As part of the detailed interpretation of the regressions, the meaning of some interactions has to be explained. Figure 6.14 does this for interactions involving X_a and Y_a as responses.

The interactive effect on motoric development at 4 years, X_a, of motoric handicap at 2 years, DX_b, and the actual score for this developmental aspect at 2 years, X_b, is shown in Figure 6.14a. This is one way of capturing the nonlinear dependence present; for children with a motoric handicap the developmental stage at 4 years is

Table 6.8 *Selected regression equations for variables on development of 350 children.*

Response	Explanatory variables					
Y_a	$Y_b.Z_c$	Y_b	Z_c	DY_b	P	*const.*
est. reg. coeff.	.091	.479	.056	−.400	.070	−.207
st. error	.022	.038	.030	.081	.035	
ratio, t	4.2	12.5	1.8	−4.9	2.0	
Y_b	Y_c	U_c	W_c	DX_c	S	*const.*
est. reg. coeff.	.211	−.112	−.118	−.242	−.207	−.058
st. error	.052	.055	.033	.098	.058	
ratio, t	4.0	−2.0	−3.6	−2.5	−3.6	
Y_c	O	S	*const.*			
est. reg. coeff.	−.125	−.226	−.143			
st. error	.038	.055				
ratio, t	−3.3	−4.1				
X_a	$X_b.DX_b$	X_b	DX_b	Y_b	*const.*	
est. reg. coeff.	.597	1.034	1.125	.303	1.374	
st. error	.170	.173	.502	.055		
ratio, t	3.5	6.0	2.2	5.5		
X_b	$X_c.O$	X_c	O	$S.DX_c$	S	DX_c
est. reg. coeff.	.121	.218	−.136	−.246	−.356	.001
st. error	.037	.066	.038	.091	.092	.111
ratio, t	3.3	3.3	−3.5	−2.7	−3.9	.0
X_b, continued	DY_c	*const.*				
est. reg. coeff.	−.231	−.118				
st. error	.091					
ratio, t	−2.6					
X_c	O	*const.*				
est. reg. coeff.	−.206	−.051				
st. error	.032					
ratio, t	−6.4					

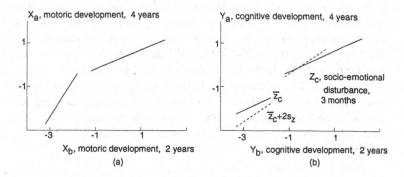

Figure 6.14 *Plots to interpret interactive effects on some repsonses in the child development data.*
Interaction (a) $DX_b * X_b$ for response X_a; (b) $Y_b * Z_c$ for response, Y_a. Observed mean and standard deviation of a variable Z_c denoted by $\bar{z}_c$ and s.

closer to that at 2 years whereas the previous developmental stage is a poorer predictor otherwise, that is the regression slope is then less steep.

The interactive effect on cognitive development at 4 years, Y_a, of cognitive development at 2 years, Y_b, and socio-emotional disturbance at 3 months, Z_c, is shown in Figure 6.14b. It is of a similar kind in the sense that better prediction of scores at 4 years is possible from those at 2 years for children with high socio-emotional disturbances at 3 months than otherwise; the dashed-line slope for highly disturbed children (scores two standard deviations above the mean $\bar{Z}_c$) is steeper than for children with average scores on Z_c. However, there is an additional interpretation, due to the fact that almost no children have strong socio-emotional disturbances at 3 months and very good cognitive development at age 2 years, i.e. the predicted dashed line above values of 1 for Y_b is remote from the observations. Thus, there are essentially strong additive effects of socio-emotional disturbance at 3 months and cognitive development at 2 years for handicapped children (values of Y_b below -1.5), but essentially no additional effect for children with a good cognitive development at 2 years.

Figures 6.15 and 6.16 give the graphical summaries of the independence structure in some of these relations, set out in a number of sections for simplicity of presentation. From the selected models some of the main conclusions are as follows. At least one of the risk factors at birth is directly or indirectly related to the different developmental aspects, no matter whether the development at 3 months, 2 or 4 years is considered. When the 4-year responses are related directly to the baseline variables, that is omitting the intermediate variables, relatively weak dependencies are found, showing that most of the influence of the baseline variables is exerted via the intermediate responses at 2 years and 3 months. The importance of the variables O, S used in designing the study is, however, confirmed.

If the information at 2 years is used for predicting development at 4 years, then the risk factors known at birth and the information available at 3 months are no longer directly relevant, the information on the same aspect measured at the earlier stage of 2 years becoming the most important predictor. Other variables recorded as potentially explanatory improve prediction considerably in addition to information on previous developmental stage of the same aspect.

The lack of direct dependence of psychic disorder, A_a on A_b, and of A_b on A_c indicates that the physician's judgement of psychic disorder does not lead to measurements which are stable over time. There are many directly explanatory variables for socio-emotional development at 4 years, Z_a, which nevertheless explain no more than 37% of the variability in the response. This is likely to reflect that more important explanatory variables for this developmental aspect have not been recorded; one possible such variable is integration into the kindergarten group.

The above conclusions are based on a long series of separate regression analyses. In addition, we have examined the correlation matrices of the residuals of the three quantitative variables Y, X, Z from their respective regression equations at each time point. The residual correlation for variable pair (Y, X) is always positive, those for (Y, Z) and (X, Z) negative and the sizes of the correlations decrease with time.

There are two consequences. Firstly, the correlations might be interpreted as suggesting a hidden explanatory variable; thereby graphs in the observed variables are interpreted as summary graphs of the kind studied in Section 8.5. Secondly, there is a loss of statistical efficiency in fitting the equations by separate least-squares analyses. The residual correlations are, however, sufficiently small

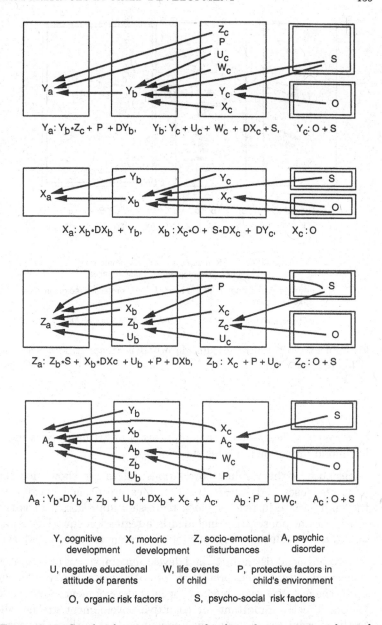

$Y_a: Y_b*Z_c + P + DY_b,$ $Y_b: Y_c + U_c + W_c + DX_c + S,$ $Y_c: O + S$

$X_a: X_b*DX_b + Y_b,$ $X_b: X_c*O + S*DX_c + DY_c,$ $X_c: O$

$Z_a: Z_b*S + X_b*DXc + U_b + P + DXb,$ $Z_b: X_c + P + U_c,$ $Z_c: O + S$

$A_a: Y_b*DY_b + Z_b + U_b + DX_b + X_c + A_c,$ $A_b: P + DW_c,$ $A_c: O + S$

Y, cognitive development	X, motoric development	Z, socio-emotional disturbances	A, psychic disorder

U, negative educational attitude of parents   W, life events of child   P, protective factors in child's environment

O, organic risk factors   S, psycho-social risk factors

Figure 6.15 *Graphical representation of independence structure for each response and its repeated measures in turn.*
Lines within boxes not displayed; equations for Y and X in Table 6.8.

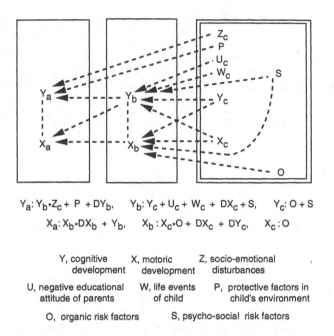

$$Y_a: Y_b \cdot Z_c + P + DY_b, \quad Y_b: Y_c + U_c + W_c + DX_c + S, \quad Y_c: O + S$$

$$X_a: X_b \cdot DX_b + Y_b, \quad X_b: X_c \cdot O + DX_c + DY_c, \quad X_c: O$$

Y, cognitive development	X, motoric development	Z, socio-emotional disturbances
U, negative educational attitude of parents	W, life events of child	P, protective factors in child's environment

O, organic risk factors   S, psycho-social risk factors

Figure 6.16 *Graphical representation of independence structure for two of the responses jointly.*

that we considered the extra complication of fitting by maximum likelihood unjustified, at this stage at least.

Because of the residual correlations, edges are shown for the joint responses in Figure 6.16. The lines are dashed because of the multivariate instead of block regression approach, that is the other joint response is not included in a regression equation. Some qualitative conclusions from the joint representation are as follows.

Psycho-social risk at birth and developmental stages at 3 months remain important direct predictors for both aspects of development up to the age of 2 years but not for later development. Additional factors directly important for cognitive development are not directly relevant for motoric development so that with an improvement in some of these factors improved cognitive development can be expected irrespective of stage of motoric development.

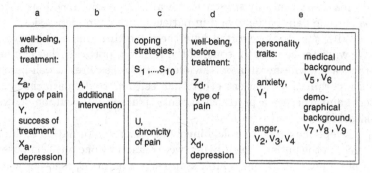

Figure 6.17 *First ordering of types of variable for chronic pain patients. Patient's well-being three weeks after stationary treatment (box a) and before (box d); an intervention (box b) intermediate for the former, response to the latter; stage of chronic pain, U, and coping strategies S_i are stacked to display hypothesized independence given pretreatment well-being (box d), personality characteristics, medical and demographic background variables (box e); double-lined box indicates relations taken as given.*

6.5 Treatment of chronic pain

Chronic pain means a considerable reduction of quality of life for a patient, stressful situations for his family and at work as well as financial burdens if in spite of treatment no lasting improvement can be achieved.

Important goals in research about chronic pain are therefore to identify and detect early risk factors that pain will become chronic and to develop treatments and interventions which many chronic pain patients will judge to be effective.

In previous investigations at the chronic pain clinic in Mainz a score for chronicity of pain had been developed which since then has proved to be a stable predictor of treatment success. The score is different from but associated with duration of pain. It incorporates four broad, rather distinct aspects of a patient's type of pain: frequency and intensity of pain, localization of pain, medication usage, history of pain treatment. In addition, there is evidence that certain strategies to cope with pain gain importance with respect to successful treatment, when chronicity is high, that is when high scores measuring stage of chronic pain have been reached. Separately, special psychotherapeutic interventions have been shown to decrease substantially tendencies towards depressive reactions both in chronic pain patients and in patients with acute depression.

With the study reported here there are three main objectives. We want to see whether some of the results reported earlier can be replicated, whether it is worthwhile to study measures of well-being of a chronic pain patient other than self-reported treatment success and whether a psychotherapeutic placebo intervention shows sizeable effects.

Telephone contacts, solely involving general information and interest in the patient's well-being, were offered by one physician over a period of three months after stationary treatment, together with the option for the patient of contacting this physician again. The patient's well-being is measured three months after the patient has left the hospital in terms of the patient's judgement of success of the stationary treatment, Y, the reported typical type of pain, Z_a and a depression score, X_a.

In a previous investigation it had been hypothesized that certain strategies to cope with pain $S_1, \ldots, S_{10}$ are independent of chronicity of pain, U, given pretreatment conditions, personality characteristics as well as medical and demographic information about the patient. This, together with our knowledge about potential relevant explanatory variables, leads to the hypothesized chain order shown in Figure 6.17. There are two stacked boxes to capture the hypothesized independence to be tested; two of the measurements of well-being are available three weeks after stationary treatment, X_a, Z_a, and before treatment, X_d, Z_d.

In Table 6.9 properties of marginal distributions from earlier investigations are contrasted for some chronic pain patients and some acute pain patients. The 83 chronic patients had head- and backache. The 43 patients with acute pain had to undergo knee surgery, many of them being athletes. Some of the main distinctions result directly from this grouping. Acute pain patients have on the average a lower chronicity score, U, they are younger, V_7, report fewer years with pain, V_5, and fewer serious illnesses, V_6.

Explanations for other observed differences are less direct, for instance, that the chronic pain patients tend to be more depressive, X, more anxious, V_1, but are completely comparable to the acute pain patients with respect to expression of anger, V_2, V_3, V_4. Quite compatible with the observed differences in personality characteristics are differences in some of the strategies to cope with pain. Among chronic pain patients there is for instance a higher ten-

Table 6.9 *Summaries of marginal distributions for variables of previous studies of chronic and acute pain patients.*

	Variable	chronic; head, back; $n = 83$		acute; knee $n = 43$	
		mean	st. dev.	mean	st. dev.
Y,	treatment success	14.3	(9.51)	23.8	(10.78)
S_1,	minimizing relevance	11.7	(4.65)	13.6	(3.77)
S_2,	relative optimism	9.7	(4.59)	11.6	(5.03)
S_3,	distraction	12.6	(4.83)	14.1	(4.76)
S_4,	self-control	16.2	(4.73)	16.8	(4.17)
S_5,	positive thinking	16.3	(4.65)	18.3	(4.37)
S_6,	preoccupation	10.6	(5.23)	8.9	(4.82)
S_7,	social retreat	9.2	(5.92)	5.7	(4.19)
S_8,	resignation	8.6	(5.14)	5.2	(3.88)
S_9,	self-pity	9.9	(5.18)	8.4	(5.45)
S_{10},	aggressive reactions	5.8	(4.05)	4.4	(3.69)
U,	chronicity of pain	5.0	(1.18)	2.6	(0.96)
X,	depression	19.6	(9.09)	11.3	(8.62)
V_1	trait anxiety	42.2	(10.65)	35.6	(7.54)
V_2,	anger supression	20.1	(3.51)	20.9	(3.74)
V_3,	anger expression	15.6	(4.34)	15.4	(3.84)
V_4,	anger control	26.8	(6.04)	28.4	(6.52)
V_5,	years of pain	11.2	(10.00)	2.3	(4.01)
V_6,	serious illnesses	4.3	(2.85)	2.7	(2.08)
V_7,	age	50.2	(13.30)	37.3	(12.14)
V_8,	gender, female	67%		58%	
V_9,	schooling, general	65%		40%	
	medium level	23%		42%	
	higher level	12%		19%	

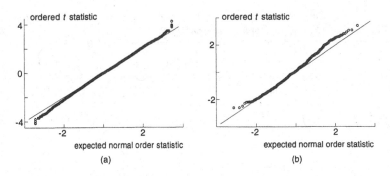

Figure 6.18 *Normal probability plot of t statistics of cross-product terms in trivariate regressions for variables on chronic pain.*
(a) Plot for all variables. (b) Plot for stage of chronic pain, U, coping strategies S_i and their potential explanatory variables; each cross-product includes U.

dency to social retreat, S_7, and resignation, S_8, and less utilization of relative optimism, S_3, and positive thinking, S_5. The reported treatment success, Y, is much higher for acute than for chronic pain patients.

Consistency with these properties of chronic pain patients is well confirmed by the typical values for patients in the present study, displayed in Table 6.10. The very similar typical values for the patients with the placebo intervention, A $(n = 28)$, and without $(n = 57)$ show, in addition, that two groups with and without additional intervention are quite comparable with respect to characteristics before treatment. The stronger check on group comparability, with logistic regression of intervention as response on main effects and interactions considered one at a time, gave no evidence to the contrary.

A further comparison of typical values for the 85 patients analysed with typical values for those patients who had incomplete records and were excluded from our analysis gave no indications of possible selection bias, the largest t statistics having absolute value 1.73; patients with complete records had a slightly higher average score for self-pity.

The checks for nonlinear relations in trivariate marginal distributions are summarized in Figure 6.18. In general most t statistics, even if they are considerably larger than 2, lie along the line

Table 6.10 *Summaries of marginal distributions; variables of present study with 85 chronic pain patients.*

Variable		hotline $n = 28$		no hotline $n = 57$	
		mean	st. dev.	mean	st. dev.
Y,	treatment success	10.1	(6.90)	8.5	(5.66)
X_a,	depression	17.2	(9.48)	23.2	(12.01)
Z_a,	typical pain intens.	5.8	(2.71)	6.5	(2.22)
A,	intervention, yes	100%		0.0%	
S_1,	minimizing relevance	10.4	(4.24)	10.9	(4.87)
S_2,	relative optimism	10.5	(4.05)	10.1	(4.60)
S_3,	distraction	12.3	(3.91)	12.3	(4.35)
S_4,	self-control	16.4	(3.79)	15.7	(4.32)
S_5,	positive thinking	15.5	(4.05)	15.1	(5.40)
S_6,	preoccupation	10.3	(4.00)	11.4	(6.00)
S_7,	social retreat	11.0	(6.14)	10.3	(6.45)
S_8,	resignation	9.9	(5.38)	10.6	(4.95)
S_9,	self-pity	10.9	(4.50)	12.1	(5.83)
S_{10},	aggressive reactions	7.5	(4.88)	7.6	(4.45)
U,	chronicity of pain	5.3	(1.38)	4.8	(1.40)
X_d,	depression	22.8	(9.59)	24.4	(9.66)
Z_d,	typical pain intens.	6.7	(1.96)	7.2	(1.88)
V_1	trait anxiety	45.2	(7.59)	47.3	(11.69)
V_2,	anger supression	21.0	(3.52)	22.7	(4.04)
V_3,	anger expression	16.9	(5.39)	16.6	(3.95)
V_4,	anger control	25.5	(5.23)	27.6	(6.03)
V_5,	years of pain	10.0	(11.30)	9.3	(9.76)
V_6,	serious illnesses	7.4	(4.17)	6.0	(4.20)
V_7,	age	50.6	(13.19)	48.0	(13.40)
V_8,	gender, female	75%		61%	
V_9,	schooling, general	54%		51%	
	medium level	32%		26%	
	higher level	14%		23%	

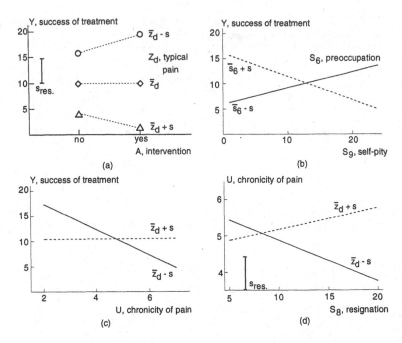

Figure 6.19 *Plots to interpret interactive effects on two responses in treatment of chronic pain data.*
a) to c) Response is Y. d) Response is U. Standard deviation of residuals in the regression denoted by $s_{res.}$; observed mean and standard deviation of a variable X denoted by $\bar{x}$, s, respectively.

of unit slope in Figure 6.18a and may therefore be interpreted as being well compatible with random variation. Some of the points deviating clearly from the unit line involve the responses of main interest. One suggests an interactive effect of preoccupation with pain, S_6, and self-pity, S_9, on treatment success, Y_a, and, indeed an interactive effect of these variables remains sizeable when other important predictors have been selected, as is reflected in regression equation for Y_a in Table 6.11 and in the plots of interaction effects on this response shown in Figure 6.19.

Chances of reporting good treatment success are about equally bad for patients who are not allowing themselves to feel self-pity and do not think intensively about pain as for patients who are doing both extensively; effects are shown in Figure 6.19b for patients

Table 6.11 *Selected regressions of responses* Y_a, X_a, Z_a.

Response	Explanatory variables					
Y_a	A	S_6	S_9	$S_6.S_9$	Z_d	$A.Z_d$
est. reg. coeff.	13.27	.94	.73	−.07	−3.15	−1.80
st. error	4.15	.25	.22	.02	1.14	.58
ratio, t	3.2	3.8	3.4	−4.6	−2.8	−3.1
Y_a, continued	U	$U.Z_d$	V_9	*const.*		
est. reg. coeff.	−5.94	.67	2.57	24.16		
st. error	1.59	.22	.68			
ratio, t	−3.7	3.1	3.8			
X_a	A	S_6	S_9	$S_6.S_9$	S_3	Z_d
est. reg. coeff.	−35.26	−1.45	-1.75	.12	−3.30	−5.69
st. error	10.27	.46	.45	.03	1.00	1.89
ratio, t	−3.4	−3.1	−3.9	4.0	−3.3	−3.0
X_a, continued	$S_3.Z_d$	X_b	V_1	V_5	$A.V_5$	*const.*
est. reg. coeff.	.44	.28	.39	−.24	1.21	64.91
st. error	.13	.14	.15	.21	.39	
ratio, t	3.3	2.1	2.6	−1.1	3.1	
Z_a	A	S_4	$A.S_4$	S_5	S_6	$A.S_6$
est. reg. coeff.	−2.35	.05	.30	−.17	.12	−.31
st. error	2.13	.08	.11	.07	.05	.09
ratio, t	−1.1	.6	2.7	−2.6	2.6	−3.3
Z_a, continued	U	Z_d	V_1	V_9	*const.*	
est. reg. coeff.	.36	.48	−.09	−.68	7.01	
st. error	.141	.115	.024	.264		
ratio, t	2.6	4.1	−3.6	−2.6		
Z_a	A	S_5	S_6	$A.S_6$	U	Z_d
est. reg. coeff.	−1.57	−.10	.13	−.34	.36	.33
st. error	1.81	.05	.04	.09	.14	.14
ratio, t	−.9	−2.2	3.0	−3.7	2.5	2.5
Z_a, continued	$A.Z_d$	V_1	V_9	*const.*		
est. reg. coeff.	.65	−.08	−.60	7.15		
st. error	.21	.02	.26			
ratio, t	3.1	−3.4	−2.3			

Table 6.12 *Correlations for responses* Y, X_a, Z_a. *Marginal correlations among residuals from their regressions; partial correlations of residuals with squares of residuals given main effects.*

| | \multicolumn{3}{c}{marginal corr.} | | | \multicolumn{3}{c}{residual corr.} | | | \multicolumn{3}{c}{partial, res. squared} | | |
	Y	X_a	Z_a	Y	X_a	Z_a	Y^2	X_a^2	Z_a^2
Y	1			1			−	.18	.26
X_a	−.45	1		−.28	1		.00	−	−.29
Z_a	−.72	.34	1	−.53	.31	1	−.13	−.26	−

having been offered phone contacts and having other variables at mean level. For patients with higher typical pain before treatment, prediction of treatment success is not further improved by knowing the chronicity of pain, while for patients with lower typical pain, reported success of treatment is better the lower the chronicity of pain; see Figure 6.19c. Finally, the additional placebo intervention has no or even a worsening effect on reported success of treatment for patients with average or high typical pain before treatment; patients with lower typical pain report better treatment success if they have been offered the possibility of further phone contacts after stationary treatments than otherwise; see Figure 6.19a.

To estimate these three interactions 9 degrees of freedom are used. This is a lot compared to the available number of patients, $n = 85$. Similarly, the regressions for the other two responses shown in Table 6.11 are fairly complex relative to the sample size. Thus, even though they point to possible important explanatory variables which explain more than 55% of the variability of each response, there is a danger of overinterpretation. To avoid this it is best to try to replicate the results in a study with more patients. Then, it might also be possible to decide which of the two alternative regression equations for Z_a is to be preferred. These have most explanatory variables in common but differ in $A * S_4$ or $A * Z_d$ as the important additional interactive effect.

The additional intervention is important for each of the three responses, the particular effect type depending on specific charac-teristics or coping strategies of the patient; see also Table 6.11.

Table 6.12 shows that correlations between treatment success, Y, and the other two responses, X_a, Z_a, are considerably reduced in size after the regressions, but that residual correlations remain

Table 6.13 *Correlations of chronicity of pain, U, with coping strategies, S_i. Marginal correlations (upper row); partial correlations given main effects of remaining potential explanatory variables (lower row).*

S_1	S_2	S_3	S_4	S_5	S_6	S_7	S_8	S_9	S_{10}
$-.01$	.01	.10	.08	.06	$-.03$	.11	.02	.11	.09
.05	$-.04$	.15	$-.04$	.11	$-.15$	.03	$-.14$	$-.13$	.07

substantial and that there is some evidence for nonlinear relations among residuals. Thus, in a future study with more patients efficient estimation of these relations would be worthwhile.

Chronicity of pain is linearly independent of all coping strategies, both, in the present study and in a previous study. It is reflected in the marginal and partial correlations given background variables $V_1, \ldots, V_9$ and well-being of the patient before treatment, X_d, Z_d, in Table 6.13 all being very small. There is no evidence for a dependency in the marginal trivariate distributions involving chronicity of pain, see Figure 6.16b. Nevertheless, the hypothesis of U being independent of coping strategies, $S_1, \ldots S_{10}$, is to be rejected in the present set of data, because of a strong interactive effect of resignation, S_8, and typical pain before treatment, Z_d, on chronicity of pain. In the multivariate regression denoted by

$$U : X_d + V_2 + V_3 + V_4 + V_6 + V_1 * V_7 + S_8 * Z_d$$

44.8% of the variability in chronicity is explained with 11 degrees of freedom, whereas in the regression obtained after removing $S_8 * Z_d$ only 36.3% remains as variation explained, a highly significant drop. A reanalysis of data sets from previous studies, which include information on these eight explanatory variables, could confirm whether this result is likely to be of substantive importance.

6.6 Bibliographic notes

Section 6.2 For a detailed description of the variables on glucose control, see Kohlmann *et al.* (1991). The special plots to check for nonlinearities, based on normal order statistics (see Cox and Hinkley 1974, p.470), were proposed by Cox and Wermuth (1994c). More powerful tests for checking symmetry of a distribution have been suggested by Edwards (1995, Appendix C).

Section 6.3 A description of the study design and of basic results of the cohort study is contained in Giesen *et al.* (1981). Some previous analyses with chain-graph models are reported by Streit (1996).

Section 6.4 For a detailed description of the study design and for further references, see Laucht, Esser and Schmidt (1992).

Section 6.5 Previous results on treatment of chronic pain patients and many relevant references are given by Hardt (1995) and Leber (1996).

CHAPTER 7

Some strategical aspects

7.1 Preliminaries

In Section 1.9 we discussed a simple default strategy for use when subject-matter knowledge and experience of similar sets of data suggest a natural starting point for the analysis. There are a number of reasons why such a strategy is inapplicable to the kind of example described in Chapter 6. Some of these reasons are concerned with the practicality of handling complex sequences of analysis and others are broader conceptual issues.

1. Because the studies are observational the strong balance required for the standard analyses of variance will not be achieved. While comparable linear and generalized linear models can often be set up, the orthogonality required to obtain, for example, interaction sums of squares or log-likelihoods regardless of other parameters fitted is not available so that it is in practice usually necessary to examine interactions by augmentation of the fitting of simpler models.

2. The absence of randomization in treatment assignment means that models are largely specific representations of the variation present rather than implied by the design.

3. A key role is played by hypotheses of conditional independence. These are essentially hypotheses of zero main effects and interactions in some form of regression analysis or of zero two-factor and higher-order interactions in some joint distribution.

4. Because of the complexity of the structures involved, emphasis is placed initially on regression-type models with main effects, interactions being added to models where essential for interpretation and also as additional explanatory variables to examine adequacy.

5. The controlling for possible confounding effects by conditioning and the discussion of possible biases can be crucial, whereas in

randomized experiments this is usually a relatively minor aspect connected with precision improvement.

6. Quite commonly in observational studies in the social sciences there is no background quantitative theory to guide model formulation. Absence of quantitative theory does not, however, mean absence of important qualitative knowledge.

7. A further feature that can arise in randomized experiments but does so more frequently in observational studies is the classification of variables as response, intermediate or explanatory. It is then required to build up a chain of relations, in part suggested by substantive arguments and in part determined during the analysis, leading to the arrangement of variables in a chain of boxes. This allows a sequence of univariate analyses for each box with a single response and some form of multivariate analysis for several variables displayed as joint responses within a box.

We begin with a quick review of the main phases of analysis and then discuss a broad approach in a little more detail.

7.2 Phases of analysis

Typically we can distinguish five phases of analysis after the main research questions have been settled.

First we check data quality by looking for possible errors in the coding of variables, for outliers, for missing values and associated possible selection effects. We do this, in particular, by producing overall summary statistics for each variable alone. For quantitative variables these statistics include the mean, the standard deviation, smallest and largest values, selected percentiles and sometimes stem and leaf displays; for qualitative variables we give just relative frequencies, typically as percentages.

By means of probability plots of summary statistics, we search for indications of strong nonlinear effects or interactive effects. These often point to relations of substantive interest but sometimes also to outliers not detectable from one-dimensional analyses.

Where there are missing values an attempt is made to complete records from original sources, to replace isolated missing values by reasonably imputed values or, at least, to understand the reason for the missing values and to document comparability of selected groups or to describe systematic selection effects that may affect

the comparisons of primary interest. At the end of this first phase of
analysis the set of data for subsequent detailed study is determined.

A first ordering on substantive grounds of the variables as responses, intermediate variables and purely explanatory variables
determines an order for further analysis. In a second phase pairwise associations are documented. Marginal associations may be
reflected in correlations, in mean differences with standard errors
or in log-odds again with standard errors or, sometimes, more com-
pactly in ratios corresponding to studentized regression coefficients
in linear or logistic regressions. These may be contrasted where possible with corresponding partial measures. The purpose is to gain
qualitative insight into relevant conditional associations.

In a third phase the relevant regression equations are developed.
With any fairly large number of potentially important variables
complete backward selection schemes are not feasible. A combination of main effect regressions, forward selection for strong interactive effects followed by backward selection for stable strong effects
has proved to be a satisfactory strategy. Lists of studentized regression coefficients computed as if they were included next into a
selected equation provide some reassurance that important further
effects have not been overlooked.

When two different regression equations provide acceptable explanations for the same response either both equations are reported
or a synthesis is obtained by combining both sets of explanatory
variables, possibly reducing the resulting equation again by backward selection.

In a fourth phase the set of agreed regression models is described
with the aid of a graphical representation, the regression coefficients and their standard errors set out and, where necessary their
meaning explained in the way described in Section 4.5. The aim is
both a quantitative summary of the analysis and a semi-qualitative
substantive interpretation.

Any of the phases may need to be repeated in the light of information uncovered later in the process. The whole process is likely
to be iterative.

7.3 An outline strategy

We now discuss in a little more detail a broad strategy for addressing situations such as those described in Chapter 6. A delicate balance is involved between the need to specify an approach

sufficiently precisely that extreme arbitrariness is avoided and in-dependent workers can understand the route adopted and, on the other hand, the avoidance of extreme rigidity so that special informal considerations specific to particular problems can be taken into account.

We make virtually no use of automatic variable selection procedures partly because they force arbitrary choices between models that fit equally well. Nevertheless, as a component to the analysis we need a broad approach to multiple regression, including logistic regression or some generalization, for use when the number of explanatory variables is large, so that some simplification by omitting variables is needed and at the same time the possibility of some nonlinear effects, including interactions, can be recognized. We call the following approach *model development*, described in the following for a single response.

First we specify those explanatory variables that should be included in all equations fitted, for example because the variables represent treatment effects of central concern or because they are known to have important effects. Next we regress the response linearly on all explanatory variables including also any nonlinear terms that on general grounds are likely to be required. Thirdly, by examining studentized regression coefficients, i.e. regression coefficients divided by their standard errors, we eliminate small terms selectively until one or more simple well-fitting models are achieved. Then we examine, one term at a time, the effect of adding squared terms and cross-product terms into the model. Any clearly significant nonlinear terms uncovered in this way need detailed interpretation before we decide how to incorporate them into the final model. To check that no important effect has been overlooked a last step is to list the contributions of the omitted primary explanatory variables if added back into the regression one at a time. The resulting procedure thus has elements of simplification by omitting variables and of augmentation by uncovering nonlinearities and interactions.

The most delicate part of the procedure is often the third step. In principle, if there are several well-fitting models of roughly equal complexity we want to isolate them all, yet fitting of all possible regressions with a given number of explanatory variables is typically not feasible unless the number of such variables is quite small.

Typically interpretation will require the variables to be ordered as responses, intermediate and explanatory in the light of subject-matter considerations, as in the specific applications described in Chapter 6.

Suppose for simplicity of exposition there are just three vector random variables, Y_a, a set of responses, X_b, a set of intermediate variables and U_c, a set of purely explanatory variables. A broad approach to the analysis is as follows.

1. For each component of Y_a separately, develop in the way outlined above a regression model for the dependence on the baseline explanatory variable U_c. Where two components of Y_a are of a very similar nature, use common components of U_c in the two regressions, at least initially.

2. For each component of Y_a separately, develop also a model for the regression on X_b and U_c. For all components of U_c in the model for step 1 but not selected in the new model examine the regression coefficients for compatible size and sign.

3. Either for all X_b or for all components selected at step 2 regress on U_c. Hence examine how the effects of U_c on Y_a found in step 1 may be interpreted partly by X_b being an intermediate variable.

4. For continuous components of Y_a and X_b calculate residual covariance and concentration matrices. Consider whether some maximum likelihood fitting of multivariate models is needed to supplement the component by component analyses. In particular, some combination of multivariate regression and multivariate block regression may be needed to describe appropriately the dependencies. Consider the possible usefulness of derived response variables, as discussed later in Section 8.2.

5. Test any substantive research hypotheses of conditional independence suggested a priori.

6. Conditional independencies suggested by empirical analysis may be incorporated into the interpretation; in principle, the possibility of alternative simplifying interpretations has to be considered. Block regression results may sometimes suggest redefinition of the status of the joint response by separating variables into stacked boxes. Sometimes, even the total separation of a block into stacked boxes may result, i.e. each box containing just a single variable. Introduction of some further ordered blocks may result after a combined evaluation of prior knowledge about the variables and the dependencies judged important by analysis; see, for example, Section 6.3.

7. Provided the resulting structure is not confusingly complicated, combine into a single graphical representation of the dependencies, with associated models described primarily via regression coefficients. Here it may sometimes be best to begin with separate graphs for each component of Y_a.

8. Test for consistency with the composite structure proposed, for example by a likelihood ratio test split into components as illustrated in Section 8.4.

An important special case arises when some components of Y_a are new observations on properties that have been measured at one or more earlier times and included as components of X_b and possibly of U_c. In such situations of repeated measurements it will normally be wise when considering a component in Y_a as response to take the corresponding components at earlier times as variables to be included in the regression equation.

7.4 Role of parsimony

In complex systems the number and types of possible independencies are very large. Simplification and idealization are essential if understanding, especially qualitative understanding, is to be achieved. In the main types of application studied in this book, conditional independence is regarded as an especially insightful simplification and therefore we aim for models that, while reasonably consistent with the data and with subject-matter considerations, contain appreciable numbers of conditional independencies. This can be regarded as a special instance of the search for parsimonious description, in particular description with the number of unknown parameters kept as small as feasible.

This is an important and fruitful principle but has to be applied with caution. For example, if the primary objective of a study is to assess the effect of particular explanatory variables on particular responses, then a parameter or parameters for those effects should be estimated, with a suitable standard error, even if the estimated effects are small and statistically insignificant. More generally, we are not recommending uncritical use of a strategy of setting to zero all parameters with insignificant estimates. Even when one or more parsimonious models with appreciable numbers of conditional independencies are chosen as a basis for the main interpretation, it may be wise in reporting on the data to include estimates of parameters that have been set to zero in the primary model. This extra information may, in particular, be needed later, for example to compare with future related investigations.

In the theory of empirical prediction equations, it is shown that, in principle, rather than setting small regression coefficients to zero it may be preferable to 'shrink' the small coefficients towards zero or some other theoretically special value. This is not a route we follow here, in particular because it would entirely fail to achieve the desired goal of qualitatively understandable interpretation. However, particularly with relatively small amounts of data, all simplifications such as conditional independency are, of course, provisional.

7.5 More formal issues

7.5.1 General remarks

While the process of finding a basis for interpretation will contain elements, often substantial, of qualitative judgement, it is useful to formalize these as far as feasible. This can be done in various ways which we now briefly discuss, although some we shall not use in this book.

We need to distinguish two quite different situations. In the first we have a number of models all of which have the same agreed target, for example the prediction of some new value or the estimation of a parameter that can be defined to have the same substantive interpretation in all models. Then model choice as such has no direct substantive interpretation and the reasonable approach may often be to make convenient cautious assumptions supplemented by formal or informal sensitivity analysis to check on the stability of the conclusions.

The more interesting possibility is where the different models have qualitatively different interpretations. Here the goal is to find one model or several models from those under consideration that are reasonably consistent with the data.

Some routes to model selection are as follows.

1. Formal procedures may be developed for choosing one out of a number of candidate models, often nested in sequence, corresponding, for example, to a degree of a polynomial, the order of an autoregression and so on.

2. Posterior probabilities may be calculated or approximated for

each of a given set of models. This yields incisive comparison of the relative effectiveness of two or more models but meaningful assignment of the prior probabilities of the models and of the conditional prior densities of parameters within the models is in our view rarely feasible in the contexts we are considering. Also the possibility that none of the models is satisfactory may escape detection if this approach is followed on its own.

3. Tests of goodness of fit may be applied designed to be sensitive against fairly specific alternatives. Any evidence against the model under test will then be combined with implications for model improvement.

4. Broad tests of goodness of fit designed to test against a broad range of alternatives may be used; they have the disadvantages of poor sensitivity for detecting specific kinds of departure and of not giving direct suggestions for model improvement.

We next discuss the indirectly related notion of what in econometrics is called model encompassing.

7.5.2 Predicting the results of fitting other models

An aspect of model choice somewhat connected with goodness of fit is model encompassing, a concept that has been suggested in particular in econometric contexts. Essentially the idea is that a 'correct' model should be able to predict the result of fitting an 'incorrect' model to the same data. In particular, by seeing whether such predictions are correct, model adequacy can be tested. In the more interesting case where the models are not nested there is a connection with the study of tests of hypotheses where the null and alternative hypotheses are of different kinds.

More broadly, the need to predict from one analysis what the main results of some other reasonable analysis would have been arises, in particular, when reanalysis of results reported in the literature is not feasible. Implications of generating processes to be discussed in Section 8.5 show one such possibility for systems with only single responses. Another simple example of predicting regression results can arise with a binary response variable if the results of a probit regression are presented. If later it is required to compare with the results of logistic regression of some other data the conversion will usually be easily accomplished because, except in the tails, the two models are virtually equivalent. In general, however, the translation will not be so simple.

7.6 Formal techniques

In the present context automatic choice of one out of a number of models seems sensible only when that choice is contained within some larger, more complex analysis, for example the representation of multiple time series via vector autoregressions for prediction and control or the fitting of curves by polynomials for use in calibration or as an intermediate step in analysis. Otherwise it is more sensible to use methods that summarize evidence about model choice so that, in particular, essentially arbitrary choices between virtually equally well-fitting models are avoided.

Where several sets of similar data are under analysis it will usually be best to fit a single type of model for each set. In particular, the use of formal model choice procedures to fit lines to some sets, parabolas to others and cubic curves to yet others is potentially confusing especially if the degree chosen is used as a basis for substantive interpretation.

We may be considering k different models with log-likelihood functions

$$l_1(\theta_1), \ldots, l_k(\theta_k),$$

with parameters of dimensionalities $d_1, \ldots, d_k$. If we want measures of the relative merits of the different models in representing the data it is reasonable to use the log-likelihood functions and, at least as an approximation, the maximized values $\hat{l}_1, \ldots, \hat{l}_k$.

If, say, model 1 is nested within model 2, i.e. the models are specified in such a way that model 1 can be recovered from model 2 when certain parameters are set to zero, the likelihood ratio test statistic $2(\hat{l}_2 - \hat{l}_1)$ can be used to test consistency with the simpler model, using the chi-squared distribution with $d_2 - d_1$ degrees of freedom.

The object here is rather different, in some sense to provide a more immediate pointer to model selection. The quantities $\hat{l}_j$ themselves cannot be used directly for this because the more complicated model would always be favoured. Some penalty for extra parameters is needed pointing towards

$$l_{c,j} = \hat{l}_j - cd_j,$$

for a suitable constant c. We see no compelling arguments for particular choices of c. Any such argument would require quantitative specification not only of knowledge comparing the different models, but also of knowledge about the parameter values within each model. Values such as $c = 2$ and $c = \frac{1}{2} \log n$ have been proposed. The latter has the qualitatively appealing feature of penalizing extra parameters more strongly in large samples, where small effects may be statistically significant but perhaps substantively unimportant.

A different and simpler argument is to set $c = \frac{1}{2}$, this being a rough correction for the bias in adding unnecessary parameters. Then a successful model and all models within which it is nested will have about the same value of the index and one would normally look at the simpler such possibilities.

These issues are most pressing in large samples when, as just noted, quite small departures from a given model may be highly significant. Note, however, that calculations of precision based on oversimplified models of error structure will often seriously underestimate the real error in large studies.

Mechanical use of model selection procedures such as $l_{c,j}$ is, in our view, to be avoided. Departures from a model should be reported if statistically significant and based on realistic assessment of errors. The departures may then be judged to be too small to be of subject-matter importance, at least for the time being, but it seems to us that this has to be regarded as an explicit subject-matter judgement not to be hidden in some automatic procedure.

7.7 Significance tests compared with other criteria

Significance tests examine consistency with a model and therefore, when there are just two models, are in principle suggesting that the data are consistent with either, both or none of the two models. In many contexts this seems an advantage over procedures that confine themselves to comparing two models.

The use of significance tests to choose between a sequence of nested models, for example to choose the degree of a polynomial, seems unwise unless there are a priori reasons for choosing low-degree models. The best procedure of the significance test type in such problems is probably to choose a value p_0 for the maximum degree likely to be used, and to test consistency with a polynomial of degree p via a comparison of the mean square of deviations with $p_0 - p$ degrees of freedom with the residual mean square from the maximal model. This will usually lead to a lower confidence limit on the 'true' value of p. If this were to be used as a selection device a quite low confidence coefficient may be appropriate, even 0.5.

As noted in Section 7.5, the distinction between significance tests and other bases for guiding model choice is particularly marked with large data sets where some of the model choice procedures, applied mechanically, are quite likely to lead to models inconsistent with the data, as judged by a significance test. Of course with very large data sets there is also the issue of whether the assessment of precision is sound. Sources of variability that are unimportant in small studies may be dominant in large ones.

7.8 Tests of goodness of fit

This is not the place to review the extensive literature on significance tests, their usefulness and their limitations. Two related criteria for classifying such tests are as between

1. tests that are sensitive against departures in specified directions, versus those with some sensitivity against broad classes of departure. Such procedures are often quite closely tied to the estimation of a parameter used to amplify the initial model, although tests of separate hypotheses need not be considered in that way.

2. tests which, when evidence against the null hypothesis is found, give some guidance as to the nature of the departure, versus those which do not. Global tests against saturated models are often of the second type, an example being likelihood ratio tests to test agreement between a sample covariance matrix and that implied by a linear structural model with latent variables. Here 'rejection' of the null hypothesis is not accompanied by any direct indication of which part of the complex model is defective and this is a limitation of this type of procedure.

7.9 Multiple use of tests

The analysis of complex data involves in effect the study of many questions simultaneously and therefore the use of many different components of analysis. Further, the development empirically of relatively simple models that fit the data again involves typically a chain of analysis. The interpretation of any one of these tests or confidence interval calculations is in terms of hypothetical long-run frequencies of error, but what are the properties of a complex chain of such procedures?

The relative neglect of formal multiple comparison issues in our discussion here is based on the following considerations.

Firstly, where a number of distinct issues are being addressed it is usually best to attach a measure of uncertainty to each issue separately, leading to a direct use of standard tests and intervals.

Secondly, where interest lies in an overall simplified model based, for example, on a number of conditional independencies, it is desirable to apply an overall test of consistency with this model and then to specify also all other models of about the same complexity that are consistent with the data at one or more significance levels. Then, if there is a simple model that is a good representation of the underlying process generating the data it will be included in the confidence set of 'true' models with a probability of error specified by the level used in the test of adequacy, no allowance for selection being used. Where there are several such models consistent with the data any choice between them has to be on external subject-matter grounds. These considerations obviate the need for multiple comparison adjustments, at least in principle, although the full specification of alternative models can be very difficult in complex problems.

Thirdly, if the issue at stake is not so much the justification of simplifications, but rather the establishment of nonzero effects, protection against selection effects is important, especially whenever an effect is reported solely because it is the largest of a number of effects examined. Some protection is achieved, for example, via the use of an overall test, such as an F test or a likelihood ratio test of a global null hypothesis. or less formally via the probability plotting techniques used in Chapter 6.

The issue of overinterpretation based on selecting large effects for study is important in some fields where data are encountered that are largely without structure, but does not appear a pressing matter in the applications reported here.

7.10 Conclusion

In conclusion, we re-emphasize the broad points listed in Section 7.1. In many ways the most crucial aspects of model choice and formulation are that the right questions should be asked and links with subject-matter considerations established.

7.11 Bibliographic notes

Sections 7.2, 7.3. Miller (1990) provides a strong warning of the dangers of uncritical use of variable selection procedures in regression. The particular procedure recommended here is, in detail, new. Cox and Snell (1974) give further principles for selecting variables. The role of tests of nonlinearity is discussed by Cox and Wermuth (1994c); see also Cox and Small (1978).

Section 7.5. For discussion of Bayesian model selection procedures, see Spiegelhalter and Smith (1982).

Section 7.6. There is a very extensive literature warning against overemphasis on and overinterpretation of significance tests, going back at least to Yates (1951). For a classification of various kinds of significance test, see Cox (1977) and, in a much more expository mode, Cox (1982).

CHAPTER 8

Some more specialized topics

8.1 Preliminaries

In this final chapter we take up a number of matters that quite commonly arise in applications. First we deal in turn with the analysis of variables that are derived from the initial primary measurements in a study and then with the related issues of hidden or latent variables and errors of measurement and the connection with the large literature on linear structural models. We then develop further the use of graphical representations addressing the problem of deducing the implications of a given model for a new set of variables formed by marginalizing and conditioning. The important complication of missing values is then discussed. Finally, we return to the broad topic, touched on in Section 2.12, of the role of notions of causality in interpreting data arising in the types of study with which we are dealing in this book.

8.2 Derived variables

The initial analysis of a body of data typically starts from a set of variables, which we call *raw variables*, that have been recorded in a way set out in the specification of the data collection process. Often, however, some of the variables most suitable for interpretation are combinations of raw variables and, as mentioned previously, we call these derived variables.

Such variables can be of various kinds, one broad distinction being between those that are based on an a priori calculation or judgement and those that are determined from the data under immediate analysis. The former are the more common.

Among the former are simple combinations of observations of the same type, such as the difference between some variable after an intervention and its value before the intervention. Note, however, that differences may not be the most sensitive combination to take. Also, especially in observational studies, it will be wise to examine the baseline distributions of the variable concerned. If ratios seem more appropriate than differences, it will usually be better to work with differences of logarithms.

Illustration. In a study of the effect of a policy intervention on the recidivism rate of young offenders, a variable for analysis would be the difference for each region between the log-odds of recidivism within a given time, for example one year after the intervention and before the intervention. The interpretation of such differences, especially in observational studies, is hazardous, and adjustment by regression for other changes may be necessary. For instance, the availability of data on control regions where no policy change is implemented will strengthen interpretation.   □

A second possibility is that sometimes particular combinations of variables of different kinds are conventionally used to capture certain features of the combination. For example, Quetelet's index, weight divided by the square of height, is a conventional measure of obesity intended to be independent of height. A further common possibility is that the data for each individual contain repeat measurements of some quantity at more or less regular time intervals and the focus of interest is the comparison of individuals rather than the detailed structure of the variation within an individual. Then it will often be sensible to calculate for each individual and to analyse as derived variables a number of summary measures. Examples of such summary measures are means, measures of dispersion, and quantities connected with trend or cyclical behaviour.

When multiple measures of response are available on each individual we can distinguish two broad situations. In the first the individual component responses have a strong meaning and it is likely that the primary interpretation will be component by component. By contrast, it may be that derived combinations of the component responses provide the best basis for understanding. Sometimes both approaches are useful.

Illustration. If blood pressure is a final response variable of interest, both diastolic and systolic components would normally be available. Because of the wide understanding of the nature of the two components and of the normal practice in the literature of reporting them separately, it would usually be necessary to report conclusions on the two components separately, or in some contexts concentrate on one component. In some ways a more ef-

fective approach, however, may be to look for some combination
of the two components, in particular a simple linear function of
the log-components, that will express the conclusions in a concise
and understandable way. This would be an instance of a derived
variable determined from the data. Later in this section we discuss in more detail an investigation in which simple combinations
of free fatty acid concentrations seem more interpretable than the
separate concentrations of individual fatty acids.   □

The most widely known technique for determining derived variables is discriminant analysis and its direct generalization, canonical correlation analysis. Here one finds, for a particular contrast
or dependence of interest, that combination of the component response variables that captures the dependence of interest most
sensitively. When discriminant analysis is used for interpretation
rather than as a rule for classifying new individuals it will usually
be wise to replace the estimated discriminant function by a simple,
more easily interpreted version.

Illustration. A linear discriminant function based on log diastolic and log systolic blood pressure might have coefficients that
were nearly equal or nearly equal with opposite signs. In that
case it would, for interpretation usually be sensible to use numer-
ically equal weights corresponding respectively to the geometric
mean and the ratio of the two original components. More generally, when dealing with nonnegative quantities measured in comparable units, use of linear combinations of logs will enable simple
combinations to be recognized; sometimes, when standard units of
mass, length and time are involved considerations of dimensional
analysis, i.e. the determination of combinations invariant under
changes of units, are helpful.   □

We now outline a technique useful when a number of component
response variables, Y, are related to a set of explanatory variables,
X, and the individual explanatory variables are regarded as of intrinsic interest, but the possibility of using derived response variables is to be examined. We look for combinations Y_j^* of responses
Y related only to single components X_j of X.

In effect each derived variable is a discriminant function for
linear regression on a component of X. Put differently, Y_1^*, say, is
conditionally independent of all remaining components of X once
X_1 is given; see Figure 8.1.

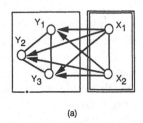   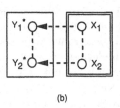

(a)   (b)

Figure 8.1 *Graphical representation of independence structure for derived variables.*
(a) Graph of original variables is complete. (b) Graph for derived variables each Y^* component depends only on one X component.

The discussion is simplest if both x and Y have the same number, p, of components. Then, if the population covariance matrices are known, we have that

$$Y^* = \Sigma_{xx}\Sigma_{yx}^{-1}Y.$$

In applying this formula, sample estimates are substituted for population values and it is assumed that Σ_{yx} is singular. If Σ_{yx} is nonsingular or severely ill-conditioned it will often be possible to reformulate the requirements in a reduced number of dimensions.

If the number, q, of explanatory variables exceeds the number, p, of response variables the specified objective cannot be met and some modified target must be set, for example one involving at most p of the explanatory variables. If, on the other hand, $p > q$ conceptually the simplest approach is to use a canonical correlation analysis to find and reject $p - q$ variables which are uncorrelated with all components of x and to check that all p canonical correlations are indeed nonzero. That gives assurance that there is enough dependence present to estimate a full set of p components Y^*. Then we apply the above result for $p = q$. It can be shown that this is equivalent to the direct calculation of

$$Y^* = \Sigma_{xx}(\Sigma_{xy}\Sigma_{yy}^{-1}\Sigma_{yx})^{-1}\Sigma_{xy}\Sigma_{yy}^{-1}Y.$$

Illustration. Immediately before an operation the log concentrations of free fatty acids were measured, namely palmitic acid, Y_1, linoleic acid, Y_2, and oleic acid, Y_3. For the present purpose these are regarded as response variables. We consider just two explana-

Table 8.1 *Marginal correlations among pre-operation variables for 40 patients.*

Variable	Y_1	Y_2	Y_3	X_1	X_2
Y_1, log palmitic acid	1				
Y_2, log linoleic acid	0.90	1			
Y_3, log oleic acid	0.95	0.92	1		
X_1, blood sugar	−0.25	−0.27	−0.32	1	
X_2, gender	0.28	0.43	0.23	−0.03	1

tory variables, namely X_1, blood sugar the morning before the operation, and X_2, gender, scored as females 1, males −1. Here no linear combination of the explanatory variables is of interest, that is their individual identities are to be preserved. Table 8.1 shows the sample correlation matrix.

The derived variables are

$$Y_1^* = 110Y_1 - 18Y_2 - 164Y_3, \quad Y_2^* = -3Y_1 + 8Y_2 - 10Y_3.$$

The coefficients can be changed by constant multiples. A more detailed analysis shows that the simpler versions

$$Y_1 - Y_3, \quad Y_2 - Y_3$$

are within sampling limits. That is, the ratio of palmitic to oleic acid appears related to blood sugar level regardless of gender and the ratio of linoleic to oleic acid is correspondingly associated with gender. A further set of patients gave similar variables. Of course an ultimately satisfactory justification of these two derived variables could come only from biochemical considerations.   □

As already remarked, in using this method it is essential that the dependence between the two sets of variables is substantial and complex enough to allow a full set of derived variables to be reasonably estimated. This can be checked by verifying that the number of nonzero canonical correlation coefficients between Y and X is equal to the number of components of X.

8.3 Hidden variables

There is a long tradition in some areas of the social sciences, notably psychometrics, in interpreting multivariate data via hidden or latent variables, i.e. variables not directly observed. Factor analysis, where the aim is to uncover underlying variables from the internal structure of a set of data, is probably the best known method. The extensive literature on linear structural relations is a development in which subject-matter information and hypotheses are introduced.

In terms of the ideas discussed in this book, consideration of hidden variables may arise in three ways. We shall here give only an outline discussion.

One possibility we have already briefly discussed in Chapter 2 and will examine more fully in Section 8.5. The dependency structure present in some types of graph may be explained by supposing that the distribution of the observed variables may usefully be interpreted as arising after marginalizing and/or conditioning on hidden variables. See especially Sections 2.5 and 8.5. This is an important issue especially if the hidden variables can plausibly be interpreted substantively.

The second possibility is that it may be helpful to regard the observed value of some variable as deviating from some underlying true value by a random measurement or sampling error. For continuous variables we usually write

$$Y_m = Y_t + \epsilon,$$

where Y_t and Y_m denote respectively true and measured values and ϵ is an error of zero mean and independent of Y_t. Of course other specifications, for example involving multiplicative errors, may be required.

Illustration. In some studies involving the effects of radiation a linear dose–response relation may be expected but, especially if a wide range of dose levels is involved, errors of measurement may have standard deviation proportional to dose. (A log transformation of dose will not be appropriate if a nonzero response is expected at zero dose.) Even without any selective bias in the measurement of dose the effect of proportional measurement errors is to flatten the dose–response curve at the upper end and in that sense to underestimate the true effect.   □

For nominal variables errors are ones of misclassification, and for binary variables the simplest specification is

$$P(Y_m = 1 \mid Y_t = 1) = 1 - \alpha_1,$$
$$P(Y_m = 0 \mid Y_t = 0) = 1 - \alpha_0,$$

where the α's are misclassification probabilities. If the α's can be estimated, preferably from blind remeasurement, these equations form the basis for adjusting an observed relation for misclassification. Again more complex specification may be needed in which the misclassification probabilities are not constant.

For continuous variables, measurement error of the above simple structure in the response merely increases the residual mean square from a fitted regression relation and correspondingly lowers the precision of estimated regression coefficients. Measurement error in explanatory variables has a more complicated effect, especially in multiple regression problems. Sometimes the shape of the relation is distorted, the magnitude of regression on component explanatory variables that have substantial measurement errors is decreased and, in some ways most seriously, there is overemphasis on components that are measured precisely and correlated with important variables that are measured imprecisely.

Thus, in particular, a spuriously strong conditional dependence can be introduced on a variable that has small measurement error. Corrections for these effects can be introduced provided that the variances of the error components can be estimated via suitable replication.

The third type of hidden variable arises when a number of related measurements are available all aimed at measuring some underlying concept, such as psychological well-being. A number of instruments, i.e. questionnaires, self-assessed and otherwise, may be used, all purporting to measure this concept and all doing so imperfectly. An attempt may then be made to interpret the data not directly in terms of the observed test scores but rather in terms of a hidden variable giving the 'true' well-being to which the observations are indirectly related. Some strong and at best partially testable assumptions are involved in doing this, assumptions partly about the relations between observed and hidden features and partly about the relation between the hidden features and relevant response and possible further explanatory variables.

A rather simple instance illustrating the arguments involved is as follows. Suppose that $X_1, \ldots, X_m$ are test scores all presumed to be related to a hidden variable U which is the 'true' underlying value of the feature under study. Then, if relations are linear and the X_j deviate from U by independent random errors, we have the simplest one-factor model of Section 3.5, namely

$$X_j = \beta_j U + \epsilon_j.$$

Here the X_j and U are measured as deviations from their means and $\epsilon_1, \ldots, \epsilon_m$ are independent errors with variances $\sigma_1^2, \ldots, \sigma_m^2$. Without loss of generality we may standardize the variance of the hidden variable, i.e. take $\text{var}(U)$ to be one.

Now

$$\text{cov}(X_j, X_k) = \beta_j \beta_k \ (j \neq k),$$

so that the tetrad conditions of Section 3.5 are satisfied. Under normal-theory assumptions, the model can be fitted by maximum likelihood using one of a number of standard packages. Goodness of fit can be examined informally by comparing observed and fitted covariances or more formally by a likelihood ratio test of this special model versus a more general alternative. If that alternative allows the covariance matrix to be arbitrary a test based on a chi-squared distribution with $\frac{1}{2}m(m-3)$ degrees of freedom is obtained. If the fit is unsatisfactory one route to possible explanation is to introduce a second hidden variable, that is to use a two-factor model.

Note that in all cases where such models are fitted there is the possibility that the estimated parameters do not correspond to a legitimate covariance matrix, the so-called *Heywood cases*. This difficulty does not arise if, for instance, the partial correlations between the variables under study are substantial and all positive and if the marginal correlations are close to fulfilling the tetrad conditions.

Suppose now that there is a response variable Y thought to depend on the feature represented by the hidden variable U. If U indeed captures the underlying feature entirely then Y will be conditionally independent of $X_1, \ldots, X_m$ given U. See Figure 8.2.

The model cannot be tested directly because U is unobserved but it can be tested indirectly by noting that we now have a single-factor model in $m + 1$ variables, which can again be fitted by maximum likelihood.

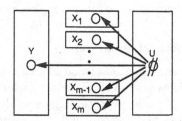

Figure 8.2 *Representation of hidden variable model.*
It has m *indicator variables for a single hidden variable* U *and one observed response* Y.

This leads us to consider three estimated covariance matrices for the full set of $m + 1$ variables $Y, X_1, \ldots, X_m$:

1. the unrestricted sample covariance matrix, $\hat{\Sigma}$;

2. the matrix $\hat{\Sigma}_{F(X)}$ in which a single-factor model is fitted to the X's and entries involving Y are equal to their observed values as in 1;

3. the matrix $\hat{\Sigma}_{F(X,Y)}$ corresponding to a single-factor model fitted to all $m + 1$ variables.

The maximized log-likelihoods are proportional to the corresponding determinants. Thus from a sample of n independent individuals we construct two likelihood ratio test statistics, written for convenience as twice the difference of maximized log-likelihoods and hence under the relevant null hypotheses having approximately chi-squared distributions. The test statistics are

$$n(\log \det \hat{\Sigma} - \log \det \hat{\Sigma}_{F(X)})$$

with degrees of freedom $\frac{1}{2}m(m - 2)$ and

$$n(\log \det \hat{\Sigma}_{F(X)} - \log \det \hat{\Sigma}_{F(X,Y)})$$

with degrees of freedom $m - 1$. The statistics test respectively consistency of the X's with a single-factor model and, given that, the consistency of the response variable Y being conditionally independent of the X's given the same hidden variable, U. The sum of the two statistics tests the overall consistency of the $m + 1$ variables with a single-factor model. The partitioning of such statistics into meaningful components distinguishing internal and external lack of fit is an important aid to interpretation which seems, however, not to be widely used.

Similar arguments apply to more complex models.

8.4 Linear structural equations

Linear structural equations developed as an extension of path analysis, factor analysis and work in econometrics on simultaneous equation models. A main purpose of such models is to represent relations between several response and explanatory variables, some of which may be hidden and some of which may mutually influence one another via relations with correlated residuals. Their typical form may be regarded as derived from a set of deterministic simultaneous linear equations for a number of variables, including possibly hidden variables, by replacing the zeros on the right-hand sides of the equations by random variables of zero mean.

The resulting models form a rich class, often nonnested in the sense that setting a term in an equation to zero may not result in an estimable submodel. No general necessary and sufficient conditions are known under which structural equation models with hidden variables have estimable parameters, i.e. have parameters all of which can be estimated precisely given a sufficiently large set of data.

Whenever the residuals are correlated with some of the variables in the equation then the interpretation of the equation parameters as measures of conditional dependence, i.e. as regression coefficients, is lost. This can be seen in very simple form from the following example. Suppose that two random variables X, Y are specified by the linear structural equations

$$Y - \gamma X = \epsilon_y, \quad X - \gamma Y = \epsilon_x,$$

where ϵ_y, ϵ_x are random terms of zero mean and equal variance. By symmetry, X and Y have equal variance, τ^2, say, and therefore the regression coefficients of Y on X and of X on Y are equal. Even if the error terms ϵ_y, ϵ_x are assumed uncorrelated the common regression coefficient is not equal to γ but rather to $2\gamma/(1+\gamma^2)$, as can be shown directly from the defining equations by noting that

$$0 = E(\epsilon_y \epsilon_x) = \text{cov}(X, Y)(1 + \gamma^2) - 2\gamma\tau^2.$$

Thus even in this simple example the direct interpretation of the coefficients in the structural equations in terms of regression coefficients is lost.

Further, in graphical representations of structural equation models missing edges do not in general correspond to a conditional or unconditional independence statement. Consider for instance the following two structural equations:

$$Y + \gamma_{yx}X + \gamma_{yv}V = \epsilon_y, \quad \gamma_{xy}Y + X + \gamma_{xw}W = \epsilon_x,$$

where V, W have a preassigned distribution.

Equations such as these two are commonly represented by a graph in the literature on structural equations. Such graphs, which it is convenient to call *structural equation graphs,* differ from those used in the present book, for instance by including error terms as nodes. If the above two equations are represented in such a structural equation graph, then the two equations are regarded as in some sense determining Y and X respectively and there are missing edges between Y and W and between X and V. In general, however, Y and W are neither marginally nor conditionally independent. In particular, if ϵ_y and ϵ_x are correlated the model is saturated, as can be seen by counting parameters. That the two edges are missing serves to make the model identifiable but no independency is implied.

In some structural equation models, however, an independency interpretation is possible. For example, univariate recursive regressions with independent residuals, multivariate regressions and seemingly unrelated regressions can all be regarded as special cases of linear structural equations.

In a matrix formulation of structural equations containing Y as responses and X as purely explanatory, we may write

$$\Gamma_y Y + \Gamma_x X = \epsilon,$$

where ϵ is a vector of errors of zero mean. Then univariate recursiveness is present if the rows of Γ_y can be permuted to triangular form, whereas in a multivariate regression model Γ_y is diagonal. In the corresponding graphical representations there are no cycles from a variable back to itself and there is at most one edge for any variable pair. If, in addition, the error components are mutually independent then the model represented by the structural equation graph is equivalent to being given by a directed acyclic graph. If error components are correlated then for diagonal Γ_y the model can be characterized alternatively by a dashed-line joint-response chain graph.

Qualitative variables in structural equations are usually assumed to be generated by dividing an underlying normal distribution into two or more classes. This implies that interactive effects of two or more variables on a response cannot be represented.

Linear structural equations include factor analysis models, the simplest form of which was discussed briefly in Sections 3.5 and 8.3. They aim to condense a number of variables, often called items, into one or more summary scores. An underlying assumption is, as mentioned previously, that the observed variables are conditionally independent given the hidden variables. That is, the covariance matrix of the observations is induced by the relations between the hidden variables. Similar assumptions and implications hold for so-called latent class models, where the hidden variable takes a limited number of values rather than being continuously distributed. In both cases, i.e. in both factor analysis and latent class analysis, a merely internal analysis of the relations among items is obtained but the usefulness typically depends on representing relations to other response variables, an aspect of external analysis.

In summary, linear structural equations are a rich family of models especially for quantitative variables. They have, however, to be used with considerable care whenever primary interest attaches to conditional independencies and to the interpretation of parameters as regression coefficients, and this is why such models have received little emphasis in this book.

8.5 Implications of generating processes

8.5.1 Preliminaries

In Chapter 2 we described a number of types of graph representing independencies and dependencies among a set of variables. We put particular emphasis on joint-response chain graphs and the special case of (generalized) univariate recursive regression graphs because these are well suited to expressing substantive research hypotheses either formulated at the start of an investigation or developed during statistical analysis. We now address the general issue outlined in Section 2.6, namely that, starting with a graph in nodes V representing a potential generating process for variables in V, we may wish to examine a reduced set S conditionally on a set C. In general terms we partition V as $\{S, C, M\}$ and consider the properties of the variables in S conditionally on those in C and hence marginalizing over those in M. What relations are implied in this new system?

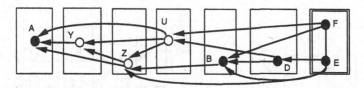

Figure 8.3 *Independence graph for subset of variables concerning university drop-out.*
Consistent with the results of Section 6.2 after marginalizing over variables X and C; see Figures 6.7 and 6.8.

With a specific set of data the properties of the distribution corresponding to S given C could be explored directly. Our interest here, however, is to say in general what is implied for the new system by a given specification for V irrespective of special nonzero numerical values of parameters.

Illustration. Suppose a model corresponding to Figure 8.3 is thought to represent a recursive process by which the data for some of the variables on university drop out can be generated, the graph being compatible with the results of Section 6.3; see especially Figures 6.7 and 6.8. If this structure holds, what relations can be expected in a different investigation

(i) in which information on some of the variables, say on grades, U, and self-judgement, Y,Z, is not available, that is if one marginalizes over some of the variables?

(ii) in which a selected subgroup of persons is studied, such as students who have repeated a high school class, D at level 1, and whose fathers have had longer formal schooling, F at level 1, that is if one conditions on special levels of some variables?

(iii) in which only a selected subgroup is studied and information on some of the variables is not provided, as in a study of only academics, A at level -1, who were high achievers during high school, U at values less than 2.5, when there is no information on a student's expected achievement, Y, and change of primary school, E, that is if there is conditioning on particular levels of variables A and U and there is marginalizing over variables Y and E?   □

We shall assume in the following that there is given for the variables under study, that is those in V, a generalized univariate

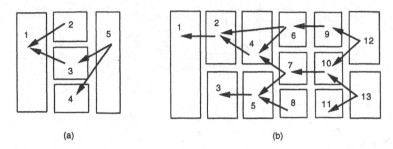

Figure 8.4 *Two illustrative generating graphs.*
Univariate recursive regression graphs (a) in 5 nodes (same as Figure 2.6a); (b) in 13 nodes.

recursive regression system as defined in Section 2.3 or a chain graph model which is independence-equivalent to such a system such as Figure 8.3. Figure 8.4 shows two further simple examples.

Given the graph for V, which we denote by G_{dag}^V, and given the partition $V = \{S, C, M\}$ we shall define in steps a new kind of graph, called a *summary graph*, from which the required properties can be obtained. First, however, we discuss briefly some distributional aspects.

8.5.2 *Some distributional aspects*

For variables $Y_1, \ldots, Y_p$ having a joint nondegenerate distribution, consider the generalized univariate recursive regression of Y_i on some of the potentially explanatory variables $Y_{i+1}, \ldots, Y_p$, $i = 1, \ldots, q$, $q = 1, \ldots, p - 1$, called directly explanatory. The regressions may for instance be linear regressions for continuous responses, (generalized) logistic or probit regressions for binary (nominal) responses. In addition to main effect terms each regression equation may contain quadratic or higher-order terms and interactive terms for some of the regressors. Any type of conditional association given the remaining directly explanatory variables has to be large enough to be of substantive interest or the involved regression coefficients are set to zero.

Whenever there is a zero regression coefficient for variable Y_j in equation i, $i < j$, this corresponds to the statement that Y_i

is conditionally independent of Y_j given the remaining potential regressors of Y_i, and by use of the separation results of Section 2.4 they are independent also conditionally given the often much smaller set of directly explanatory variables for Y_i. If there is a set of k purely explanatory variables, the distribution of which is not to be modelled, it can for the following discussion be replaced by an arbitrarily ordered set of $k - 1$ complete univariate recursive regression equations, that is by any one of the possible univariate recursive representations of their unrestricted joint distribution.

This leads to a directed acyclic graph in nodes $V = \{1, \ldots, p\}$ having attached to it boxes containing single responses, some of which may be stacked, because the response variable pair Y_i, Y_j, $j = i + 1$, is conditionally independent given $Y_{j+1}, \ldots Y_p$. In this context we name the graph G_{dag}^V a *generating graph* and take as illustrative examples the two graphs shown in Figure 8.4.

The arguments that follow are best understood as referring to linear least-squares regression equations in a multivariate normal framework. They apply also, however, to the very much broader family of problems that we shall call *quasi-linear*. In these any curvatures and higher-order interactions present are such that a vanishing linear least-squares regression coefficient implies that no dependence of substantive importance is present. That is, we exclude dependencies that are so curved or involve such special forms of high-order interaction that a corresponding measure of least-squares linear dependence would vanish.

As mentioned previously, it is important for interpretation that absence of an edge means an appropriate independence and that presence of an edge implies a dependence strong enough to be of substantive importance. To ensure the latter property in our discussion we assume the absence of a *parametric cancellation* or near-cancellation, i.e. a particular constellation of numerical values that reduces to zero a relevant combination of parameters that is in general nonzero.

A simple example in the context of linear regression is as follows. Let Y_1, Y_2, Y_3 be generated by the complete univariate recursive regression system of Figure 8.5. That is, the three defining regression coefficients $\beta_{12.3}, \beta_{13.2}, \beta_{23}$ are all substantively important so that, in particular, they are all different from zero. Now suppose that we marginalize over Y_2, i.e. we consider just Y_1, Y_3, where

$$\beta_{13} = \beta_{13.2} + \beta_{12.3}\beta_{23}.$$

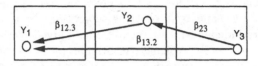

Figure 8.5 *Regression coefficients in a complete quasi-linear system for three variables.*

Now in general β_{13} will be nonzero, but it is possible that the particular numerical values are such that

$$\beta_{13.2} = -\beta_{12.3}\beta_{23}.$$

Then atypically $\beta_{13} = 0$ and we say that a parameter cancellation has occurred. Note that the cancellation has arisen within a linear system by the cancellation of effects of different paths.

Two different situations where such path cancellations cannot occur are as follows. Either the generating graph is a *tree*, that is each node has just one incoming arrow and all have a same common source node, or the relations are MTP2 (*multivariate totally positive of order 2*), that is each partial covariance given all remaining variables is zero or positive. This implies that no least-squares regression coefficient is negative, irrespective of the conditioning set. In either case the generating graph does not contain a sink-oriented V-configuration. This implies in particular that the independence structure of the fully directed graph can equivalently be represented by a concentration graph having the same skeleton.

A rather different kind of parametric cancellation can arise when particular interactions or extreme nonlinearities are present; for example it is mathematically possible for two qualitative explanatory variables to have zero main effects, i.e. display marginal independencies, but nonzero interaction in their influence on a response Y. Such possibilities are excluded by definition in linear and quasi-linear systems of only main effect regressions.

8.5.3 Simple correlation-inducing paths

Before giving some general results it is useful to discuss the consequences of conditioning and marginalizing in some simple configurations. A key role is played, as already emphasized in Chapter 2, by the configurations of Figure 8.6 in which an edge i and j is absent.

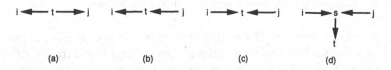

Figure 8.6 *Some simple graph configurations.*
(a) Node t is a source node; (b) t is a transition node; (c) t is a collision
node; (d) t is a descendant of collision node s.

We consider the consequences of marginalizing and conditioning on t, using the notation $\rho_{ij}, \rho_{ij.t}$ respectively for the marginal correlation between the variables represented by nodes i and j and for their conditional correlation given t.

In both Figures 8.6a and 8.6b, $Y_i \perp\!\!\!\perp Y_j \mid Y_t$ is specified so that we start from $\rho_{ij.t} = 0$ and get $\rho_{ij} = \rho_{it}\rho_{jt}$, which is nonzero because the presence of the relevant edges implies that ρ_{it} and ρ_{jt} are both nonzero. Thus $\rho_{ij} \neq 0$ and when we marginalize over t we say that an *edge between i and j is induced*. For a reason that will be explained below, we define the added edge to be a dashed line, $i\,\text{-}\text{-}\text{-}\,j$, for the source node (a) and to be an arrow pointing to i, $i\longleftarrow j$, for the transition node (b). Conditioning on node t yields independence by definition in both cases and so no new edge is added.

By contrast, for a collision node (see Figure 8.6c), we start from $Y_i \perp\!\!\!\perp Y_j$ and $\rho_{ij} = 0$; by marginalizing over the collision node no additional edge is induced, but by conditioning on it a nonzero partial correlation results

$$\rho_{ij.t} = -\rho_{it.j}\rho_{jt.i} \neq 0,$$

so that an edge between i and j must be added by conditioning on t; we use an undirected full-line edge. There is an extension of the argument to the case of Figure 8.6d, where the conditioned-upon node t is the descendant of a collision node s. Again $\rho_{ij.t} \neq 0$ and an undirected full-line edge is induced between i and j by conditioning on t.

Induced dashed or full lines will, in particular, be present in the covariance graph and the concentration graph, respectively, of the distribution considered.

8.5.4 Induced covariance and concentration graphs

A limited answer to the study of the distribution of S given C can be given via the structure of the covariance and concentration matrices, $\Sigma_{SS.C}, \Sigma_{SS.C}^{-1}$ as represented by their graphs. The simplest instance of this, i.e. of an empty conditioning set, has already been given in Section 2.3.

The *induced covariance graph given* C, $G_{\text{cov}}^{S.C}$, is an undirected dashed-line graph, again of the type described in Section 2.3, in which an edge is present between i and j if and only if $Y_i \perp\!\!\!\perp Y_j \,|\, Y_C$ is not implied by G_{dag}^V. Similarly, the *induced concentration graph given* C, $G_{\text{con}}^{S.C}$, is an undirected full-line graph, of exactly the type described in Section 2.3, in which an edge between i and j is present if and only if $Y_i \perp\!\!\!\perp Y_j \,|\, Y_{C \cup S \setminus \{i,j\}}$ is not implied by the generating graph G_{dag}^V.

In principle, the separation results of Section 2.4 can be applied to each edge in turn to decide whether it will be present in the graph. In the covariance graph of S given C there will be an edge i, j if the pair has an active path relative to C. As we have described briefly in Chapter 2, this means either that there is a collisionless path between i and j outside C in the generating graph or that such a path is generated by conditioning on C. In a concentration graph of S given C there will be an edge if there is an active path in the generating graph relative to the union of sets C and S without nodes i, j. For larger graphs computationally simpler procedures are available but will not be discussed here.

Figures 8.7 to 8.10 give some examples derived from the generating graph of Figure 8.4a. Figure 8.7 shows covariance graphs obtained after conditioning on a single node which is the common neighbour node in a V-configuration and is, respectively, a sink, a transition or a source node. The special conditioning node is set aside in a double-lined box.

The corresponding conditional concentration graphs in Figure 8.8 are also subgraphs induced by the nodes of S in the overall concentration graph G_{con}^V shown in Figure 2.6b of Section 2.3.

Figure 8.9 shows covariance graphs obtained after marginalizing over a single node, again over one of the three types of common neighbour in a directed V-configuration. These covariance graphs are also subgraphs induced by the selected subset of nodes in the overall covariance graph G_{cov}^V shown in Figure 2.6c.

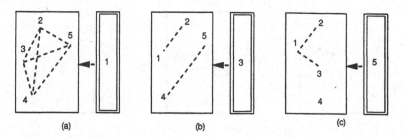

Figure 8.7 *Conditional covariance graphs obtained from the generating graph of Figure 8.4a by conditioning on different types of node.*
Conditioned (a) on the sink node 1; (b) on the transition node 3; (c) on the source node 5.

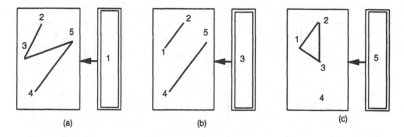

Figure 8.8 *Conditional concentration graphs obtained from the generating graph of Figure 8.4a by conditioning on different types of node.*
Conditioned (a) on the sink node 1; (b) on the transition node 3; (c) on the source node 5. All three are subgraphs of overall concentration graph of Figure 2.6b.

The corresponding concentration graphs in the selected marginal distributions are displayed in Figure 8.10.

These simple examples illustrate how different relations in a set of variables can appear depending on (a) which of the variables are considered explicitly, (b) which type of relation among each variable pair is considered, and (c) which selection effects are introduced by conditioning on variables.

As we have seen previously, undirected graphs are much less informative than fully directed graphs, both because information about the direction of effects has been suppressed and because the condensation to undirected graphs in general introduces edges ad-

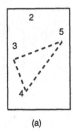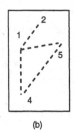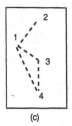

(a)   (b)   (c)

Figure 8.9 *Marginal covariance graphs obtained from the generating graph of Figure 8.4a by marginalizing over different types of node. Marginalized (a) over the sink node 1; (b) over the transition node 3; (c) over the source node 5. All three are subgraphs of overall covariance graph of Figure 2.6c.*

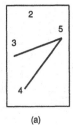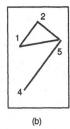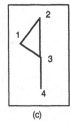

(a)   (b)   (c)

Figure 8.10 *Marginal concentration graphs obtained from the generating graph of Figure 8.4a by marginalizing over different types of node. Marginalized (a) over the sink node 1; (b) over the transition node 3; (c) over the source node 5.*

ditional to those in the graph representing a generating process. We saw in the earlier discussion that information about the direction of effects can be crucial for interpretation.

8.5.5 Simple summary graphs

To retain most of the information about directions of dependencies and all that about independencies we define a quite new kind of graph called the *summary graph* $G_{\text{sum}}^S(C, M)$ for S conditioned on C and marginalized over M. We consider first such graphs resulting after either conditioning only or after marginalizing only.

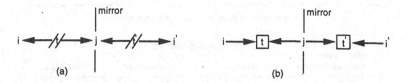

Figure 8.11 *Use of mirroring to illustrate differing implications of dashed and full lines.*
(a) Source-oriented V-configuration mirrored at node j corresponding to mirrored dashed line; implies $Y_i \perp\!\!\!\perp Y_{i'}$ and induces nonzero partial correlation $\rho_{ii'.j}$. (b) Sink-oriented V-configuration mirrored at node j corresponding to mirrored full line; implies $Y_i \perp\!\!\!\perp Y_{i'} | Y_j$ and induces nonzero marginal correlation $\rho_{ii'}$ in the conditional distribution given $Y_t, Y_{t'}$. Dependences are implied provided no parametric cancellation present.

The initial generating graph G^V_{dag} is here a graph of directed edges. Inducing new edges in general changes the character of the graph. An exception is, for instance, case (b) of Figure 8.6, i.e. of marginalizing over a transition node, where the induced directed edge can be shown to be consistent with the interpretation of G^V_{dag}. In other cases, however, the new edges are defined to be dashed- and full-line undirected edges respectively. The need to make the distinctions can be seen by mirroring a subgraph at node i, say, by replacing for example the three-node configurations of Figures 8.6a and 8.6c by the five-node configurations of Figures 8.11a and 8.11b. In the first case after marginalizing on t and t' we have $Y_i \perp\!\!\!\perp Y_{i'}$. In the second case after conditioning on t and t' we have $Y_i \perp\!\!\!\perp Y_{i'} | Y_j$. The resulting graphs are consistent with the dashed- and full-line notation used previously for covariance and concentration graphs.

Distinctive features of the three types of edge in a summary graph are further illustrated with Figure 8.12. This shows which simple univariate recursive regressions generate in the summary graph an arrow, a full line, and a dashed line connecting responses Y, X for two unjoined explanatory variables V, U, and displays the concentration and covariance graphs in the corresponding joint distributions of the four variables. In case (a), when the edge of (Y, X) is an arrow, the independencies are partly reflected in the concentration graph with $(Y, V) \perp\!\!\!\perp U | X$ and partly in the covariance graph with $V \perp\!\!\!\perp (X, U)$. In case (b), with the responses connected by a full line, all independencies show in the concentration

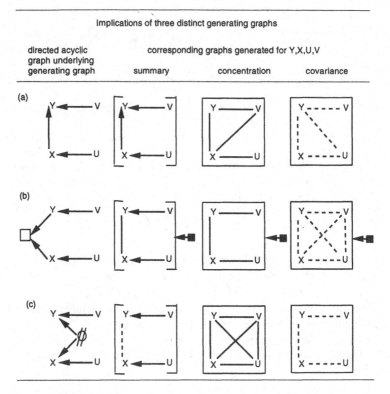

Figure 8.12 *Smallest graphs generating three types of edge connecting responses in summary graph.*
Summary graph (a) not independence-equivalent to concentration or covariance graph, (b) independence-equivalent to concentration graph, (c) independence equivalent to covariance graph.

graph and none in the covariance graph. In case (c), with a dashed line for (Y, X), the situation is reversed, the covariance graph is independence-equivalent to the summary graph and the concentration graph is complete. This emphasizes the need for two different types of undirected edge in summary graphs.

An extension of these ideas enables us to deal directly with problems in which either only a series of marginalizations is involved or only a series of conditionings. It can be seen that provided there are no parametric cancellations the effect of these operations can be only to add edges to the generating graph not to remove edges, that is to induce further dependencies but not further independencies.

If only marginalizing over the variables with nodes in a set M is involved, the effect is to modify the generating graph by adding an edge between i and j if there is a collisionless path from i to j with every node along it in M and in no other situations. The edge is an arrow pointing from j to i if j is indirectly explanatory for i and is a dashed line if i and j have a common source node along the path.

If only conditioning on the variables with nodes in the set C is involved, the generating graph is modified by adding a full-line edge between nodes i and j if i and j have a common collision node which is in C or has a descendant in C and in no other situations.

From the generating graph modified by marginalizing, the summary graph $G_{\mathrm{sum}}^{V\setminus M}(\emptyset, M)$ is obtained by deleting all nodes and edges of M. Similarly, the summary graph $G_{\mathrm{sum}}^{V\setminus C}(C, \emptyset)$ is the subgraph induced by all nodes except those conditioned on, $V \setminus C$, in the generating graph modified by conditioning only.

8.5.6 Derivation of general summary graph

We now set out the general procedure for constructing a summary graph after both conditioning and marginalizing. Starting from a generalized univariate recursive regression graph G_{dag}^{V} and a partition $V = \{S, C, M\}$, we wish to construct $G_{\mathrm{sum}}^{S}(C, M)$, a summary graph in nodes S obtained from the generating graph by conditioning on C and marginalizing over M. The graph is to retain as much information on directions as possible and all information on independencies and dependencies among the variables in S conditionally on those in C.

We can approach the task of constructing $G_{\mathrm{sum}}^{S}(C, M)$ in three ways:

(i) by marginalizing over all nodes in M to obtain $G_{\mathrm{sum}}^{V\setminus M}(\emptyset, M)$ and then conditioning on nodes in C;

(ii) by conditioning first to obtain $G_{\mathrm{sum}}^{V\setminus C}(C, \emptyset)$ and then marginalizing;

(iii) by working in arbitrary order on individual nodes and reducing the number of nodes at each step by one to obtain the summary graph of the corresponding distribution.

These must, for consistency with the laws of probability, give the same final answer. Our objective is that, if possible, we should describe and parametrize the new structure in a simple way, for example as some form of chain graph. In some cases, however, a direct reduction to such a simple form is not available, in particular because several components of the same edge may point in opposite directions.

The general rules to be followed are justified by extension of the arguments given in the previous sections. They are most conveniently described by specifying how additional marginalization and conditioning can be represented in a summary graph in which some such operations have already taken place. This implies, in particular, that they apply to the generating graph, G^V_{dag} itself. We first mark by $\diamond$ those nodes in the given summary graph $G^S_{\mathrm{sum}}(C, M)$ that have in the generating graph descendants in C.

We illustrate the procedures first on the relatively simple 6-node generating graph in Figure 8.13, then on the 13-node graph of Figure 8.4b in Figures 8.14 to 8.17.

Since edges in a path of a summary graph may be arrows, full lines or dashed lines, some of the notions introduced earlier for directed graphs have to be extended.

Nodes i and j are said to have a *collision node t in a summary graph* if t is their common neighbour in one of the following three paths

$$i \longrightarrow t \longleftarrow j, \quad i\text{---}t\longleftarrow j, \quad i\text{---}t\text{---}j.$$

More generally, they have a *collision sequence in a summary graph* if the single node t in any of the above three paths is replaced by a dashed-line path in nodes $t_1, \ldots, t_q$ for $q > 1$.

Other notions remain unchanged in a summary graph as compared with the corresponding directed generating graph. A path is collisionless if it does not contain a collision node, and a node h is a descendant of t if a direction-preserving path of arrows points from t to h.

Rules for modifying a summary graph are as follows. By marginalizing over nodes m of S the summary graph $G^S_{\mathrm{sum}}(C, M)$, an edge is added between nodes i and j if there is a collisionless path between them with every node along it in m and in no other situations. The type of edge, which may be additional to an edge of a different type already present, is determined as follows:

1. $i \longleftarrow j$ if (a) at i there is an arrowhead and at j there is an

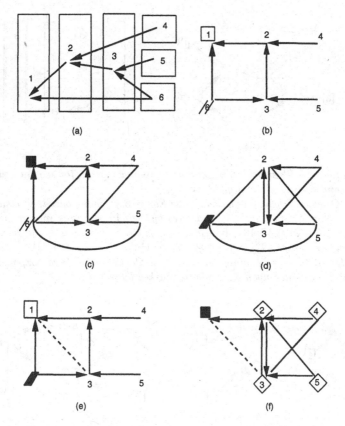

Figure 8.13 *Illustration for deriving a summary graph in different order of conditioning and marginalizing.*

(a) Generating graph of 6 nodes. (b) Underlying directed acyclic graph with node 1 marked as to be conditioned on and node 6 as to be marginalized over. (c) Graph of (b) modified by conditioning on node 1. (d) Graph of (c) modified by marginalizing over node 6 after having first deleted node 1 and its edges. (e) Graph of (b) modified by marginalizing over node 6. (f) Graph of (e) modified by conditioning on node 1 after having first deleted node 6 and its edges and marked nodes with a conditioned upon descendant by ◇. Summary graph is the subgraph induced by nodes 2,3,4,5 both in (d) and (f).

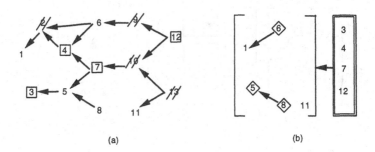

(a)   (b)

Figure 8.14 *Illustration for deriving a summary graph for several nodes in conditioning and marginalizing sets.*
(a) Directed acyclic graph underlying Figure 8.4b, with nodes to be conditioned on, $C = \{3, 4, 7, 12\}$, indicated by $\square$ and nodes to be marginalized over, $M = \{2, 9, 10, 13\}$, indicated by two lines $//$ crossing the node. (b) Summary graph $G^S_{sum}(C, M)$ for selected set $S = \{1, 6, 5, 8, 11\}$, with nodes of the conditioning set, C, set aside in double-lined box; nodes with conditioned-upon descendants marked by $\diamond$.

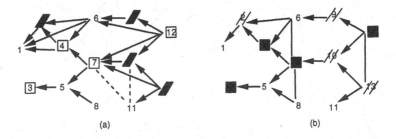

(a)   (b)

Figure 8.15 *Generating graph of Figure 8.14a modified with added edges and nodes operated on blacked in.*
Result after (a) marginalizing over the nodes of M; (b) conditioning on the nodes of C.

   arrowtail or a full line, or (b) at i there is a dashed line and at j there is a full line;

2. $i - - - j$ if at i there is an arrowhead and at j there is an arrowhead or a dashed line;

3. $i \text{———} j$ if (a) at both i and j there is a full line, if (b) at i there is a full line and at j there is an arrowtail or vice versa, i.e. if (c) at i there is an arrowtail and at j there is a full line.

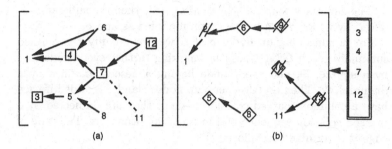

(a)   (b)

Figure 8.16 *Summary graphs corresponding to Figure 8.14a after only marginalizing over M or after only conditioning on C.*
(a) $G_{\text{sum}}^{V\backslash M}(\emptyset, M)$ *obtained as subgraph induced by all nodes except those already marginalized over, $V \setminus M$, in Figure 8.15a; with nodes still to be conditioned on indicated by* □*. (b)* $G_{\text{sum}}^{V\backslash C}(C, \emptyset)$ *obtained as subgraph induced by all nodes except those already conditioned on, $V \setminus C$, in Figure 8.15b; with nodes still to be marginalized over indicated by two lines // crossing the node and nodes with conditioned-upon descendants marked by* ◇*.*

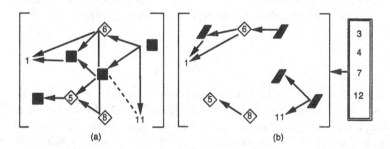

(a)   (b)

Figure 8.17 *Summary graphs of Figure 8.16 modified with added edges. (a) Result after conditioning on nodes of $C = \{3, 4, 7, 12\}$ in Figure 8.16a. (b) Result after marginalizing over nodes of $M = \{2, 9, 10, 13\}$ in Figure 8.16b. Summary graph of Figure 8.14b obtained next as subgraph induced by nodes $S = \{1, 6, 5, 8, 11\}$ in either graph.*

The summary graph $G_{\text{sum}}^{S\setminus m}(C, M \cup m)$ is then the subgraph induced by nodes $S \setminus m$ in the graph modified as above.

For conditioning on nodes c of S we first modify the marked summary graph $G_{\text{sum}}^{S}(C, M)$ by marking further every node in c by a square, $\square$, and every node having a descendant in c by a diamond, $\diamond$. Then an edge between nodes i and j is added if they have a common marked collision node or they are connected by a marked collision sequence and in no other situations. The type of edge is determined as follows:

4. $i \dashleftarrow j$ if at i there is a dashed line and at j an arrowtail;

5. $i\dashrightarrow\!\!\!\text{-}\, j$ if at both i and j there is a dashed line;

6. $i \relbar\relbar j$ if at both i and j there is an arrowtail.

The summary graph $G_{\text{sum}}^{S\setminus c}(C \cup c, M)$ is then the subgraph induced by nodes $S \setminus c$ in the graph of the nodes of S modified in the way just described.

Nodes over which it is intended to condition or marginalize are indicated as above by $\boxdot$ and $\not\!\!\bigcirc$, respectively. After the operation is completed and the graph modified by inserting any additional edges, the originating symbol is blacked in. In the next subsubsection some justification of the rules is provided. This is not needed to understand the interpretation of the summary graph set out in Section 8.5.8.

8.5.7 Explanations for induced edges

The plausibility and internal consistency of these rules is often best checked by reversing the order of marginalization. To see this it is helpful to introduce the notion of a *synthetic directed acyclic graph* which allows the independence interpretation of a given summary graph or of a subgraph thereof in terms of an associated fully directed graph.

A synthetic directed acyclic graph is obtained in several steps from a given summary graph by replacing:

(i) each dashed line $i\dashrightarrow\!\!\!\text{-}\, j$ by a source-oriented V-configuration $i \leftarrow \not\!\!\bigcirc \rightarrow j$;

(ii) each full line by a sink-oriented V-configuration $i \longrightarrow \boxtimes \longleftarrow j$;

(iii) in any directed cycle one arrow $i \dashleftarrow j$ by a path with a source-oriented V-configuration next to a sink-oriented one, that is by $i \leftarrow \not\!\!\bigcirc \rightarrow \boxtimes \longleftarrow j$ in order to break up the cycle.

The corresponding nodes are additional, need not have any substantive meaning, and therefore are called synthetic. The added configurations represent paths with nodes over which we respectively marginalize, $\cancel{\bigcirc}$, or condition, $\boxtimes$, to get back to the summary graph.

It can be shown that the independencies implied for the set of nonsynthetic nodes by the generating graph coincide with those implied by the synthetic graph.

To see, for instance, that marginalizing over t in the three-node graph $i \leftarrow t \dashdash j$ implies marginally correlated end-points, i.e. $i \dashdash j$, we replace it by $i \leftarrow t \leftarrow \cancel{\bigcirc} \rightarrow j$. Marginalizing first over t gives, as described in Section 8.5.3, a source-oriented V-configuration, $i \leftarrow \cancel{\bigcirc} \rightarrow j$, and marginalizing next over the synthetic node leads to the stated implication.

On the other hand, if $i \dashdash t \leftarrow j$ is replaced by $i \leftarrow \cancel{\bigcirc} \rightarrow t \leftarrow j$ then marginalizing first over t disconnects i and j so that no edge is induced; the end-point variables are implied to be marginally independent.

Whether the edge, induced is symmetric or is asymmetric may again be checked by mirroring. For instance, mirroring at node i gives, for a path looking like a prolonged arrow $i \leftarrow t \relbar j$, the graph

$$ j' \relbar t' \rightarrow i \leftarrow t \relbar j, $$

turning node i into a collision node, and marginalizing over i disconnects j' and j, while mirroring at node j turns j into a common explanatory variable for i and i',

$$ i \leftarrow t \relbar j \relbar t \rightarrow i', $$

so that end-points i and i' are marginally correlated. Hence the relation induced between i and j by marginalizing over t in a prolonged arrow path from j to i must be asymmetric; in fact its effects are just like those of an arrow pointing directly from j to i.

On the other hand, a symmetric relation is induced after marginalizing over t in a three-node path $i \relbar t \leftarrow j$, in which a full line is met by an arrowhead. Mirroring at node i and separately at node j leads to

$$ j' \rightarrow t' \relbar i \relbar t \leftarrow j $$

and

$$ i \relbar t \leftarrow j \rightarrow t' \relbar i'; $$

both paths are collisionless and are therefore correlation-inducing for the end-point variables after marginalizing over all nodes along them.

8.5.8 Criteria for induced relations in summary graphs

For the interpretation of summary graphs we need to develop separation criteria analogous to those for simpler graphs in Chapter 2 and, more importantly for the applications we consider, criteria for induced dependencies.

For a full interpretation of a summary graph, $G_{\text{sum}}^S(C, M)$, we consider the possibility of conditioning on a further set of nodes c contained in S. As for a directed acyclic graph, a path between i and j is called active or correlation-generating with respect to a conditioning set c if between i and j there is a path with every collision node along it either in c or having a descendant in c and every other node outside c. As mentioned previously, by contrast with a directed graph, in a summary graph there are three types of configuration with a collision node

$$i \longrightarrow t \longleftarrow j, \quad i - - - t \longleftarrow j, \quad i \longleftarrow t \longrightarrow j.$$

This means a node t in a summary graph is a collision node if and only if it is a collision node in the associated synthetic graph.

There are now two versions of an induced relation criterion, the second requiring the marking of a summary graph $G_{\text{sum}}^S(C, M)$ in the sense that every node that has a conditioned upon descendant in C has the special mark $\diamond$ attached. In both statements we assume the absence of parametric cancellations.

1. In an unmarked summary graph, Y_i is dependent on Y_j given Y_C and Y_c if and only if there is an active path between i and j relative to c.

2. In a marked summary graph modified by further conditioning on nodes c, Y_i is dependent on Y_j given Y_C and Y_c if and only if there is a collisionless path between i and j outside c.

As a first illustration we take the summary graph derived in Figure 8.13 and displayed in Figure 8.18 with just one edge missing for node pair (4,5). In this conditional distribution Y_4 and Y_5 will be dependent marginally in any quasi-linear system without parametric cancellation, since there is a collisionless path outside the empty set, for instance via the arrow pointing from 4 to 2 and the full line from 2 to 5; Y_4 and Y_5 are also conditionally dependent given

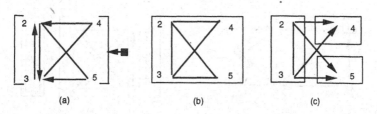

(a)   (b)   (c)

Figure 8.18 *Summary graph to Figure 8.13, with just one independence statement.*
(a) Cyclic summmary graph with $Y_4 \perp\!\!\!\perp Y_5 \mid (Y_2, Y_3)$. Independence cannot be represented by a chain graph which preserves information on Y_2, Y_3 being responses but by an independence-equivalent chain graph (b) without responses, or (c) with previously explanatory variables turned into responses.

Y_3, since there is the same collisionless path not touching node 3 from 4 to 5 via node 2. But, by conditioning on nodes 2 and 3 no further edge is induced and there is no collisionless path from 4 to 5 outside the set containing nodes 2 and 3; hence $Y_4 \perp\!\!\!\perp Y_5 \mid (Y_2, Y_3)$.

From the summary graph of Figure 8.14b it is, for instance, immediate that $(Y_1, Y_6) \perp\!\!\!\perp (Y_5, Y_8) \perp\!\!\!\perp Y_{11}$ given (Y_3, Y_4, Y_7, Y_{12}) since there are no paths connecting the three variable sets. In the summary graph of Figure 8.16a, for instance, Y_1 is marginally dependent on Y_{11} since there is a collisionless path between nodes 1 and 11 via 4 and 7; Y_1 is also dependent on Y_{11} given Y_7, since by conditioning on node 7 a collisionless path from 1 to 11 is induced passing through nodes 6 and 12.

8.5.9 Reduction of summary graph

In general the result of a combined operation of conditioning and marginalizing will lead to some pairs of nodes i and j being joined by several different types of edge simultaneously.

The most general possibility is illustrated in Figure 8.19. The arrow pointing from node 4 to node 3 is present in the generating graph, the dashed line between them is generated by marginalizing over 5, the full line by conditioning on 2 and the arrow pointing from 3 to 4 is generated by conditioning on node 1 to give $3 \longrightarrow 6 \longrightarrow 4$ and marginalizing over node 6.

It is, however, except in two cases, possible to reduce edge com-

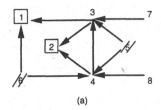 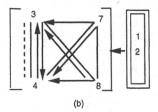

(a)   (b)

Figure 8.19 *Smallest graph generating an unreduced summary graph with one edge having 4 components.*
(a) Directed acyclic graph underlying the generating graph in 8 nodes with conditioning set $C = \{1,2\}$ and marginalizing set $M = \{5,6\}$. (b) Unreduced summary graph $G^S_{sum}(C, M)$ with $S = \{3,4,7,8\}$. Reduced graph has just a dashed line between nodes 3 and 4 and arrows pointing to them, corresponds to a saturated multivariate regression with Y_3, Y_4 as joint responses and Y_7, Y_8 being dependent joint explanatory variables.

ponents to one type without changing the independencies implied or the information on potential collision nodes. To achieve this we operate on pairs of component edges in accordance with the following rules:

(i) $\diamond \rlap{\,=}{\leftarrow} \diamond$ replaced by $\diamond \leftarrow \diamond$;

(ii) $\diamond \rlap{=}{--} \diamond$ replaced by $\diamond --- \diamond$;

(iii) $i \rlap{\,=}{\leftarrow} \diamond$ replaced by $i --- \diamond$;

(iv) $\diamond \rlap{\,=}{\leftarrow} \diamond$ replaced by $\diamond --- \diamond$.

The justification of, for instance, (iii) is seen from Figure 8.20. It can be checked from the properties we have derived that the independencies implied are the same in the two summary graphs so that an edge having an arrow and a dashed line has the same implications. Note that an arrow from U to Y is always present in a summary graph as in Figure 8.20b irrespective of it being present in the generating graph. The reason is that it becomes induced in any configuration $Y --- X \leftarrow U$ with X having a conditioned-upon descendant. The information as to which of the nodes have an incoming arrow is preserved if the reduction (iii) is applied since with a dashed line an arrow points implicitly at each response, i.e. there are arrows in a corresponding synthetic directed acyclic graph.

Figure 8.20 also shows that in some cases there are alternative

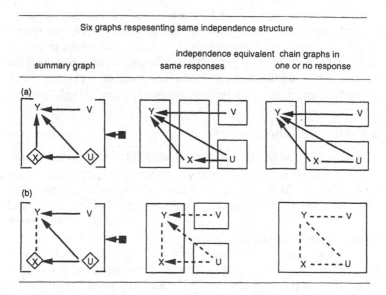

Figure 8.20 *Illustration of distinct summary graphs representing the same independence structure and compatible chain graphs.*
Just $V \perp\!\!\!\perp (X,U)$ holds. (a) Ordered responses in summary graph and two compatible generalized univariate recursive regression graphs. (b) Responses on equal footing in summary graph and two compatible graphs. being a multivariate regression chain and a covariance graph.

parametrizations in terms of chain graphs which preserve the same independencies but neglect information in the generating graph on some of the variables being responses. In some cases it is even impossible to capture the independencies in their entirety with a chain graph and to preserve information on responses in the generating graph. Figures 8.12b and 8.18 provide examples.

There are two exceptional cases where two component edges are needed: $i \mathrel{\underleftarrow{}} j$ and $\Diamond \mathrel{\underrightarrow{}\!\!\!\underleftarrow{}} \Diamond$. The first describes a relation generated by a regression and a hidden variable which has been marginalized. By the second, the independence structure given previously in Figure 8.18 can be characterized. It has exactly one independence, $Y_4 \perp\!\!\!\perp Y_5 \,|\, (Y_2, Y_3)$, and hence is independence-equivalent to a concentration graph of the same skeleton or to a directed full-edge chain graph in which Y_4, Y_5 are responses to Y_2, Y_3.

Once conditioning on joint responses has occured, originally unrelated variables explanatory for these responses become marginally related but can stay conditionally independent in the new distribution. In that case information on original directions is lost at the same time. This just reconfirms that automatic procedures for reading off an undirected graph the directions of dependence in a generating graph need not lead to substantively meaningful results.

8.6 Missing values

In the main discussions of statistical analysis in this book we have implicitly assumed that all relevant observations are available on all individuals. However, in large-scale studies missing values are very likely to occur. The detailed development of techniques to deal with the resulting complications is beyond the scope of the present book and we shall confine the discussion to some broad issues.

Sometimes values of explanatory variables are missing by design in order to economize on the costs of collecting data, but in that case balanced patterns will usually be adopted to ease analysis. We shall not consider this possibility further.

A first obvious, but nevertheless important, point is that wherever possible the data collection procedures should contain checks operating swiftly enough for missing or suspect values to be detected early and corrected.

There are two main situations to consider.

First, there may be isolated missing explanatory variables on individuals with otherwise complete data. Then, unless the proportion of individuals so affected is quite large, it is desirable to impute values for the missing entries. When the connections between variables are weak, it may be enough to replace the missing value by the mean of the observed values of that variable. In other cases regression equations may be set up to predict the missing values from observed explanatory variables with which the missing variable is strongly associated. In complex cases more elaborate imputation procedures may be needed; these are essentially generalizations of techniques introduced originally in the context of balanced experimental designs and related to the EM algorithm. If the values so imputed are treated as real observations the precision of conclusions will be overestimated and if the proportion of missing observations is appreciable some adjustment to the estimated standard errors is needed.

All these methods assume that the reason why an observation

is missing is not directly connected with the specific value that in principle should have been obtained. For example, if x_1 and x_2 are quite highly correlated explanatory variables and individuals where the value of x_2 has a large positive residual from its regression on x_1 are missing, some bias will be introduced by the above methods of imputation. Unless it is known with some confidence why the value is missing the assumptions implicitly underlying imputation are usually difficult to check effectively. It is technically possible to proceed further by specific modelling of the process leading to missing responses, but such analyses are very sensitive to the assumptions made.

The second main possibility is that primary response variables are missing for some individuals. There are a number of situations.

If some individuals have missing responses for some reason connected with the system under study it may occasionally be sensible to treat being missing as a binary response to be studied directly. Another possibility is that we may define the population under study to be responders, i.e. to study the distribution of response variables conditionally on their not being missing.

In studies in which over a substantial period intermediate response variables are measured of the same feature as the missing response, it may be reasonable to use the last observed response as a surrogate for the value of interest. Otherwise, if the response is missing at random, the individual in question should be omitted from the analysis.

If, as is often the case, the assumption of being missing at random is at best tentative, some check can be obtained by comparing the univariate and bivariate distributions of important explanatory variables in the two groups, i.e. with and without missing responses.

8.7 Causality

Except for a brief discussion in Section 2.12, we have not in this book used the words *causal* or *causality*. Yet in some sense much, if not all, scientific research has an aim of understanding the effect of one variable on another, and some notion of causality is involved in this. Our reason for caution is that it is rare that firm conclusions about causality can be drawn from one study, however carefully designed and executed, especially when that study is observational. The thrust of our discussion, however, especially the use of univariate recursive regression graphs, is to provide representations of data that are potentially causal, i.e. which are consistent with or suggestive of causal interpretation, but this is well short of actually establishing causality in a single study.

In this final section we return to these broad issues.

First it is important to distinguish a number of possible meanings of causality. The older philosophical notion that a cause is a necessary and sufficient precursor of an effect, i.e. the cause always leads to the effect and the effect never occurs without the cause, is inapplicable to the contexts we are considering. Some probabilistic notion is needed and usually some idea of multiple causation. We shall exclude from the discussion the special role of causality in quantum mechanics and also the different issue of the need for some idea of causality in classical deterministic physics.

There are essentially three interrelated senses in which causality arises in the contexts considered in this book:

1. as a statistical dependence which cannot be removed by alternative acceptable explanatory variables;

2. as the inferred consequence of some intervention in the system;

3. as the above, augmented by some understanding of a process or mechanism accounting for what is observed.

All these notions are valuable; we favour restricting the word to the final and most stringent form.

We now discuss these three meanings in turn.

For the first meaning, suppose that we have a response variable Y and that there is a partial ordering of *all possible* explanatory variables into blocks, variables in each block being responses to variables in all blocks to the right as in the specification of a generalized univariate recursive regression graph. The ordering into boxes is based on prior subject-matter considerations, often on the time sequence involved. Then, variables referring to time points before that defining X lie in boxes to the right of it. Alternatively the ordering may be based on some provisionally assumed order of explanation such as of pharmacological processes as being explanatory to a patient's overt behaviour. Then under the first definition a particular variable X is a cause of Y if Y is conditionally dependent on X given all explanatory variables in the same box as X or in boxes to the right.

Notice that this definition of causality involves all possible different explanatory variables. The implementation of the definition can, of course, use only those explanatory variables actually available, and it is the possibility that a feature unobserved, or whose existence is even unappreciated by the investigator, that is the real explanation of what is observed that makes this an ultimately unsatisfactory definition of causality. These issues are strongly connected with the modifications of independence structure induced by marginalizing and discussed in Section 8.5. Nevertheless, the search for explanatory variables that have strong contributions in regression-like representations is a crucial part of statistical analysis in many contexts.

If, however, X corresponds to a treatment variable in a randomized experiment and if there is no interaction between the treatment variable and other explanatory variables, i.e. any treatment difference is essentially uniform across all individuals under study, then any other admissible explanatory variable, observed or unobserved, affects all levels of X equally and so its effect will not bias the estimation of the effect of X.

The exclusion from the conditioning of variables that are partially responses to X is essential in this definition. Otherwise any effect of X would often be largely absorbed in further measurements made between X and Y. In the terminology of Chapter 2 the directly explanatory variables that absorb the effect of an indirectly explanatory variable are excluded so that an indirectly explanatory variable is eligible to be a cause.

Illustration. Two different groups of patients newly diagnosed as suffering from high blood pressure are studied, one receiving medication A and the other medication B, the comparison of the medications being the objective of the study. Blood pressure is available at entry into the study and after three months of treatment, and the response is the occurrence of a cardiac event within two years, regarded as a binary variable. Of course in a realistic situation many other variables would be available. Provided the initial blood pressure is measured before treatment began, in an observational study we may regress occurrence of an adverse event on initial blood pressure and treatment, examining for a possible interaction; we must consider the possibility that treatment allocation is influenced by initial blood pressure. If we find an apparent effect of treatment there remains the possibility that treatment allocation was influenced by some other variable which is the real explanation of the apparent effect. We can examine measured ex-

planatory variables for this but if we do not know how treatment allocation was determined there is always an additional uncertainty,
a gap between this first definition of causality and its realization. If
treatment allocation was randomized these ambiguities of interpretation are eliminated unless there is a strong interaction between
the treatment effect and an unmeasured explanatory variable or
there are unobserved effects intermediate between treatment and
its apparently observed effect.

The blood pressure measured after a period of treatment could
not be used in assessing the presence of a treatment effect on cardiac event rate but might be useful as an intermediate variable in
assessing whether any such treatment difference could be accounted
for by, say, improved blood pressure control.   □

The second broad approach to a definition of causality hinges
on considering actual or potential interventions in the system. For
simplicity, consider a possible cause C taking just two levels, 0,
say control, and 1, say a new treatment. For each individual we
observe C and a response Y, as well as typically further explanatory
variables. Conceptually we suppose that, for each individual j who
was observed with $C = 0$ and with response Y_{j0} we could have
had $C = 1$ and response Y_{j1}, say, this response, however, always
being unobserved. Similarly, for individuals observed with $C = 1$
we suppose that we could have had $C = 0$ with a corresponding
response, in fact unobserved. That is, for each individual there are
two responses (Y_{j0}, Y_{j1}), only one of which is observed, together
with the corresponding value of C. Such an unobserved response
is sometimes called *counterfactual*.

If there were a systematic difference between Y_{j1} and Y_{j0}, for
example if $Y_{j1} > Y_{j0}$ for all j, one could say that there is a causal
effect of C on response. A specific assumption often made is that
there is no interaction with a measured or unmeasured explanatory
variable, i.e. that

$$Y_{j1} - Y_{j0} = \Delta,$$

an assumption often called unit-treatment additivity. This can be
tested to some extent by examining for interaction between the
difference of interest and observed intrinsic explanatory variables.

This formulation is quite widely used in the context of randomized experiments where the two levels of C are assigned at random,
so that an individual who in fact had $C = 1$ indeed might well have
had $C = 0$. When the notion is applied to an observational study

it is necessary to restrict possible causes to those explanatory variables which, in the context in question, might have had different values. That is, intrinsic explanatory variables are excluded as possible causes.

Illustration. In most contexts gender is not an acceptable causal variable under this second definition. That would involve considering the question 'how would this man have responded had he been a woman?', and in most situations this would be meaningless. A possible exception is in studies of discriminatory employment practices, where the question at issue is how much would this individual have been paid given the same education, training, skills, employment history, etc., had, say, a man rather than a woman been involved.   □

A crucial issue in using this second definition concerns the conditions to be held fixed under the hypothetical change of the causal variable. The difference between responses at $C = 1$ and $C = 0$ has to be made other things being equal in some sense, which is sometimes called the *ceteris paribus* condition. If a division of the variables into ordered blocks, corresponding to a joint-response chain graph, is available, then changes in C should hold fixed variables in boxes to the right of C but the role of variables in the same box as C can be unclear. In addition, it may be necessary to consider more than one ordering of the variables, as the following illustrations show.

Illustration. In a study of the possible effect of high blood pressure on some outcome, we might have diastolic blood pressure, various biochemical measurements, for example sodium level, and demographic variables, as well as outcome measures as response variables. An arrangement in three boxes would contain (i) the outcome measures, (ii) the blood pressure, sodium level, etc., (iii) the demographic variables. We consider the hypothetical possibility that blood pressure for a patient is different from that observed and consider the consequent effect on response. It is clear that demographic variables in box (iii) should be held fixed. Whether sodium level should be held fixed under the hypothetical change of blood pressure or conceptually allowed to change depends on a hypothesis about the direction of the relation between the two. If sodium level is to be held fixed it should be included as an explanatory variable in a regression relation; if it is to be allowed to vary as in the current study, sodium level should not be used as an explanatory variable. More than one analysis may be required for a full interpretation because of uncertainties over the status of some of the variables in the causal process.   □

In randomized experiments, the notion that the treatment applied to an individual might have been different from that observed reflects the real process of design. Further, subject to unit-treatment additivity, the randomization provides protection against biases between the individuals assigned to the different treatment. Even here, however, the issue often has to be faced of what is to be held fixed as treatments vary.

Illustration. In an agricultural experiment a number of fertilizer treatments are randomly allocated to plots in some standard design. One treatment produces an exceptionally heavy crop, birds from a wide area attack these plots selectively and the final yield on them is low. The special treatment has caused a depression in yield if we regard selective attack by birds as one of the consequences of allocation to that treatment, and indeed if we formed a joint-response chain graph based on temporal ordering, attack by birds would come to the left of treatment and would be ignored in any discussion of causality. Yet it is clear, both from the view-point of understanding the data and of technological implementation, that this conclusion on its own is misleading and that an analysis in which adjustment is made for attack by birds, i.e. that variable is notionally regarded as explanatory, is needed or at least some qualification of the initial conclusions is required. Another implication of this example is the desirability of having a good choice of intermediate variables between the potential cause and the final response.   □

Illustration. Issues similar to those in the previous illustration arise over compliance and supplementary medication in connection with randomized clinical trials. That is, some patients may not follow the treatment regime to which they have been randomized, for a variety of reasons, and some patients may receive additional medication not explicitly controlled in the study protocol. Analysis by intention to treat is analogous to the direct analysis of yield in the previous illustration. The complications of noncompliance are ignored on the grounds that what is being compared are the total consequences of being randomized to one arm rather than another. While such an analysis is normally desirable, if noncompliance is appreciable either the whole investigation is suspect or further analyses will be required, especially when noncompliance can be measured and is different for the different treatments. □

The final notion of causality, most stringent and most difficult to formalize, combines the above with some understanding of an underlying process or mechanism. Of course this is hardly a precise definition; there is a wide range of possibilities between general plausibility on subject-matter grounds and full understanding based on widely accepted principles and direct evidence.

Qualitative conditions under which such interpretation becomes relatively more plausible were set out by A. Bradford Hill, and we conclude by reviewing these in slightly amended form under the following headings:

1. an effect is more likely to be causal if a firmly established subject-matter explanation of the effect and prediction of it have been established beforehand;

2. a subject-matter explanation found retrospectively does enhance the possibility of an effect being causal but is less convincing than if provided beforehand;

3. large effects are more likely to have a causal interpretation than small effects;

4. effects monotonic in some natural measure of 'dose' are more likely to be causal than those irregular with respect to dose, although in some contexts smooth nonmonotonic behaviour is to be expected;

5. an effect found repeatedly in independent studies is more likely to be causal, especially if the studies are of somewhat different form, so that a common source of bias is less likely;

6. effects showing no major interactions with many intrinsic explanatory variables are more likely to be causal;

7. massive interventions may provide information likely to have a causal interpretation;

8. effects that are specific in nature are more likely to be causal.

To use the third definition of causality given above, reasonable subject-matter understanding of a process underlying the data is required, and this calls for the first, or perhaps the second, of the above conditions to be satisfied. From one view-point, whether an

explanation is produced beforehand or later might seem imma-
terial, and indeed this is largely true when deduction from very
firmly established principles is involved. In many fields, however,
it is possible with some ingenuity to produce plausible explanations of almost any conceivable observational outcome. While it
is, of course, very important to look for explanations especially of
unexpected conclusions, due caution is usually desirable towards
explanations set up with hindsight. Ideally they should be checked
by independent study.

Massive interventions, called by Bradford Hill natural experiments, are major perturbations of a system whose behaviour following the perturbation can be studied intensively. In human populations such perturbations are usually highly unfavourable interventions, natural or otherwise.

Illustration. Much of the knowledge about the effects of large
amounts of radiation on mortality comes from intensive study of
the survivors of the Hiroshima atomic bomb, dosage being estimated from the individual's distance from the epicentre of the ex-
plosion. Alternative explanations of the enhanced mortality seem
unlikely in the extreme.   □

Illustration. Another example concerns haemophiliacs in the UK
exposed over a period to contaminated blood products. A substantial number became infected with HIV; mortality over the next 10
years among those infected was about 10 times that among the others, who retained the mortality rate holding before the exposure.
The increase was independent of the severity of the haemophilia.
The excess deaths are largely from AIDS- related causes. This
is probably the clearest available direct evidence that HIV infection causes AIDS. An alternative explanation would have to pos-
tulate an intervention affecting all, or at least a large number,
of the haemophiliacs and inducing HIV infection as a 'harmless'
side-effect. No intervention other than via the contaminated blood
products is apparent. In this case, while the process by which HIV
infection produces immunological breakdown is by no means fully
understood, there is some evidence from monkeys that adds biological plausibility to the strong empirical evidence.   □

Of the other points the last, about specificity, is probably the
most controversial and needs comment. The point is of most relevance in fields where very specific physical or biological processes

operate and is much less so where the possibly causal variables under consideration are really conglomerates of many features.

Illustrations. It is usually thought that cancers at different sites have somewhat different aetiologies, and a single explanatory variable that in an epidemiological study appeared to affect all or many cancer sites would often, although not always, be prima-facie evidence of a bias in the comparisons under study rather than evidence of a causal connection. This is a difficult issue, however, and is hard to discuss except in particular contexts.

There is strong evidence of a substantial gradient of health, as measured by many criteria, with socio-economic class, i.e. the effect is nonspecific. Quite apart from the aspect that some may be economically deprived because of their health, there is little doubt about the existence of a causal nonspecific relation. But to study this in depth would require much more detailed specification of explanatory variables, disease by disease, and isolating, for example, the effects of smoking, diet and the impact of any reduction in quality of medical care available to the poor.   □

In conclusion, while we favour restraint in making claims of causality, especially based on the analysis of single sets of observational data, the methods discussed in this book, especially the use of joint-response chain graphs, as close to fully directed as possible, are intended to develop interpretations which aim to be explanatory in as deep a sense as feasible.

8.8 Bibliographic notes

Section 8.2. For the work on derived variables, see Cox and Wermuth (1992a), Wermuth and Cox (1995) and Wermuth and Rüssmann (1993).

Section 8.3. Hidden variables underlie factor analysis (Lawley and Maxwell, 1971). Measurement error in linear models is discussed in detail by Fuller (1987) and in nonlinear models by Carroll, Ruppert and Stefanski (1995). Many of the key points are described by Cochran (1968). For a recent review with an epidemiological emphasis, see Thomas, Stram and Dwyer (1993).

Section 8.4. General algorithms for maximum likelihood estimation were introduced for linear structural equations by Jöreskog

(1981); the models are described in detail by Bollen (1989). For some critical comments on the relation with conditional independency models, see Wermuth (1992). For an early discussion of key issues, see Haavelmo (1943).

Section 8.5. The work on the implications of generating processes is new, and partly based on Wermuth, Cox and Pearl (1994). For a fast way of constructing both the covariance and the concentration graph in any conditional distribution obtained from univariate recursive regressions, see Wermuth (1995). For the concept of total positivity of distributions, see Karlin and Rinott (1980). For cyclic independence graphs and their relation to linear structural equations, see Spirtes (1995) and Koster (1996). Lack of parametric collapsibilty has been named 'faithfulness' of the graph by Spirtes, Glymour and Scheines (1993).

Section 8.6. Procedures for dealing with missing values are described in detail by Rubin (1987) and by Little and Rubin (1987). The EM algorithm for point estimation in complex problems related to ones with simple solution, and in particular for dealing with missing values, is due to Dempster, Laird and Rubin (1977).

Section 8.7. There is an extensive philosophical literature on causality, mostly but not entirely concerned with situations in which each outcome has a unique certain cause and therefore not very productive in the situations discussed in this book. For further discussion of the view-point taken here, see especially the review paper (with discussion) by Holland (1986), emphasizing the work of Rubin (1974); see also Cox (1992) and Sobel (1995). Bradford Hill (1965) discussed general criteria reinforcing a causal interpretation. For further discussion, see Rothman (1986) and, from a political science perspective, King, Keohane and Verba (1994). For discussion of statistical issues connected with discrimination in employment, see Dempster (1988), and of AIDS among haemophilic patients, see Darby *et al.* (1995).

Appendix

Tables A1 and A2 give data on 68 patients for Section 6.2.

Table A1 *Raw data on glucose control for 29 patients.*

10Y	X	Z	U	V	12W	A	B
68	35	22	32	35	37	1	1
99	38	28	30	37	105	1	-1
83	42	20	20	38	60	1	1
109	44	15	22	44	185	-1	-1
70	38	15	25	44	60	1	-1
88	41	28	35	48	204	1	-1
115	40	12	39	42	8	-1	1
99	37	21	24	42	9	-1	-1
95	30	23	31	43	204	-1	-1
70	41	19	26	45	60	1	1
135	35	21	24	45	1	-1	1
74	36	18	29	44	240	1	1
108	34	15	27	48	5	-1	1
102	25	24	31	41	145	1	-1
62	41	18	17	47	18	1	-1
68	39	19	28	37	96	-1	1
83	40	14	22	44	16	1	1
83	35	17	25	44	240	-1	-1
83	44	8	14	48	18	1	1
132	11	21	18	45	96	-1	-1
90	27	33	30	27	140	-1	1
129	16	28	26	37	84	-1	1
100	41	15	22	40	0	1	-1
69	39	22	25	38	120	-1	1
115	34	17	21	42	135	1	-1
102	42	17	25	38	120	1	1
91	28	21	29	44	288	-1	-1
97	27	30	42	46	240	-1	-1
92	24	22	17	46	216	-1	1

Table A2 *Raw data on glucose control for 39 patients.*

10Y	X	Z	U	V	12W	A	B
88	44	11	11	46	60	1	1
73	42	17	25	47	210	-1	-1
59	42	12	31	46	8	1	-1
74	40	15	14	47	195	-1	1
89	40	27	30	44	255	-1	1
64	37	14	29	34	5	1	-1
104	26	31	29	43	75	-1	-1
90	30	28	28	43	240	-1	1
141	32	15	27	47	60	-1	-1
117	42	14	23	40	84	-1	-1
105	42	15	22	44	108	1	-1
92	34	14	13	45	240	1	-1
112	41	21	24	38	140	-1	-1
100	24	13	33	39	200	-1	-1
108	21	25	34	38	25	-1	-1
76	31	21	21	39	180	-1	1
80	42	16	23	34	170	-1	1
92	28	23	28	35	75	-1	1
74	26	17	22	38	110	1	1
68	42	16	32	43	216	-1	1
85	44	14	18	41	180	1	-1
110	42	14	8	45	170	-1	1
110	37	12	14	48	60	-1	-1
54	38	13	17	40	120	1	-1
87	35	23	34	27	85	1	1
130	34	18	20	37	3	-1	1
102	26	20	30	42	120	1	-1
85	38	16	13	45	85	1	1
113	35	25	29	39	145	-1	1
136	34	20	24	38	110	1	1
80	35	19	24	40	120	1	-1
87	42	20	17	35	100	-1	-1
86	46	10	8	47	2	1	-1
117	45	18	17	38	220	1	1
107	26	22	24	37	120	-1	-1
85	37	22	27	38	270	-1	1
84	33	11	18	39	250	-1	1
78	39	25	18	36	240	1	1
101	31	24	31	47	260	-1	1

References

Agresti, A. (1984) *Analysis of ordinal categorical data.* New York: Wiley.

Agresti, A. (1990) *Categorical data analysis.* New York: Wiley.

Anderson, T.W. (1973) Asymptotically efficient estimation of covariance matrices with linear structure. *Ann. Statist.* **1**, 135–141.

Anderson, T.W. (1984) *Introduction to multivariate statistical analysis,* 2nd edn. New York: Wiley.

Andersson, S.A. and Perlman, M. (1995) Testing lattice conditional independence models. *J. Mult. Anal.* **53**, 18–38.

Atkinson, A.C. (1985) *Plots, transformations and regression.* Oxford: Oxford University Press.

Barndorff-Nielsen, O.E. and Cox, D.R. (1994) *Inference and asymptotics.* London: Chapman & Hall.

Bartlett, M.S. (1935) Contingency table interactions. *Suppl. J. Roy. Statist. Soc.* **2**, 248–252.

Berkson, J. (1946) Limitations of the application of fourfold table analysis to hospital data. *Biometrics* **2**, 47–53.

Besag, J.E. (1974) Spatial interaction and the statistical analysis of lattice systems (with discussion). *J. Roy. Statist. Soc. B* **36**, 192–236.

Birch, M. W. (1963) Maximum likelihood in three-way contingency tables. *J. Roy. Statist. Soc. B* **25**, 220–233.

Birkes, D. and Dodge, Y. (1993) *Alternative methods of regression.* New York: Wiley.

Bollen, K.A. (1989) *Structural equations with latent variables.* New York: Wiley.

Box, G.E.P. (1960). Fitting empirical data. *Ann. New York Acad. of Sci.* **86**, 792–816.

Box, G.E.P. (1966) Use and abuse of regression. *Technometrics* **8**, 625–629.

Box, G.E.P. and Jenkins, G.M. (1976). *Time series analysis,* 2nd edn. San Francisco: Holden Day.

Bradford Hill, A. (1965) The environment and disease: association or causation. *Proc. Roy. Soc. Medicine* **58**, 295–300.

Carroll, R.J., Ruppert, D. and Stefanski, L.A. (1995) *Measurement error in nonlinear models.* London: Chapman & Hall.

Chatfield, C. (1988) *Problem solving.* London: Chapman & Hall.

Cochran, W.G. (1965) The planning of observational studies of human populations. *J. Roy. Statist. Soc. A* **128**, 234–265.

Cochran, W.G. (1968) Errors of measurement in statistics. *Technometrics* **10**, 637–666.

Cook, R.G. and Weisberg, S. (1982) *Residuals and influence in regression*. London: Chapman & Hall.

Cox, D.R. (1972) The analysis of multivariate binary data. *Appl. Statist.* **21**, 113–120.

Cox, D.R. (1977) The role of significance tests (with discussion). *Scan. J. Statist.* **4**, 49–70.

Cox, D.R. (1982) Statistical significance tests. *Brit. J. Clinical Pharmac.* **14**, 325–331.

Cox, D.R. (1984) Interaction. *Int. Statist. Rev.* **52**, 1–31.

Cox, D.R. (1990) Role of models in statistical analysis. *Statist. Sci.* **5**, 169–174.

Cox, D.R. (1992) Causality: some statistical aspects. *J. Roy. Statist. Soc. A* **155**, 291–301.

Cox, D.R. and Hinkley, D.V. (1974) *Theoretical statistics*. London: Chapman & Hall.

Cox, D.R. and Small, N.J.H. (1978) Testing multivariate normality. *Biometrika* **65**, 263–272.

Cox, D.R. and Snell, E.J. (1974) The choice of variables in observational studies. *Appl. Statist.* **23**, 51–59.

Cox, D.R. and Snell, E.J. (1981) *Applied statistics*. London: Chapman & Hall.

Cox, D.R. and Snell, E.J. (1989) *Analysis of binary data*, 2nd edn. London: Chapman & Hall.

Cox, D.R. and Wermuth, N. (1990) An approximation to maximum likelihood estimates in reduced models. *Biometrika* **77**, 747–761.

Cox, D.R. and Wermuth, N. (1992a) On the calculation of derived variables in the analysis of multivariate responses. *J. Mult. Anal.* **42**, 162–170.

Cox, D.R. and Wermuth, N. (1992b) A comment on the coefficient of determination for binary responses. *Amer. Statist.* **46**, 1–4.

Cox, D.R. and Wermuth, N. (1993a) Linear dependencies represented by chain graphs (with discussion). *Statist. Sci.* **8**, 204–283.

Cox, D.R. and Wermuth, N. (1993b) Some recent work on methods for the analysis of multivariate data in the social sciences. In C.R. Rao (ed.), *Multivariate analysis: future directions*. Amsterdam: Elsevier.

Cox, D.R. and Wermuth, N. (1994a) A note on the quadratic exponential distribution. *Biometrika* **81**, 403–408.

Cox, D.R. and Wermuth, N. (1994b) Response models for mixed binary and quantitative variables. *Biometrika* **79**, 441–461.

Cox, D.R. and Wermuth, N. (1994c) Tests of linearity, multivariate normality and the adequacy of linear scores. *Appl. Statist.* **43**, 347–355.

Darby, S.C., Ewart, D.W., Giagrande, P.L.F., Dolin, P.J., Spooner, R.J.D. and Rizza, C.R. (1995) Mortality before and after HIV infection in the complete UK population of haemophiliacs. *Nature* **377**, 79–82.

Darroch, J.N. (1974) Multiplicative and additive interaction in contingency tables. *Biometrika* **61**, 207–214.

Darroch, J.N., Lauritzen, S.L. and Speed, T.P. (1980) Markov fields and log-linear models for contingency tables. *Ann. Statist.* **8**, 522–539.

Dawid, A.P. (1979) Conditional independence in statistical theory (with discussion). *J. Roy. Statist. Soc. B* **41**, 1–31.

Dempster, A.P. (1972) Covariance selection. *Biometrics* **28**, 157–175.

Dempster, A.P. (1988) Employment discrimination and statistical science (with discussion). *Statist. Sci.* **3**, 149–195.

Dempster, A.P., Laird, N. and Rubin, D.B. (1977) Maximum likelihood from incomplete data via the EM algorithm (with discussion). *J. Roy. Statist. Soc. B* **39**, 1–38.

Duncan, O.D. (1975) *Introduction to structural equation models.* New York: Academic Press.

Duncan, O.D. (1984) *Notes on social measurement, historical and critical.* New York: Sage.

Edwards, D. (1995) *Introduction to graphical modelling.* New York: Springer.

Erikson, R. and Goldthorpe, J.H. (1993) *The constant flux.* Oxford: Oxford University Press.

Fahrmeir, L. and Tutz, G. (1994) *Multivariate statistical modelling based on generalized linear models.* New York: Springer.

Fitzmaurice, G.M., Laird, N.M. and Rotnitzky, A.G. (1993) Regression models for discrete longitudinal responses (with discussion). *Statist. Sci.* **8**, 284–309.

Frydenberg, M. (1990a) Marginalization and collapsibility in graphical interaction models. *Ann. Statist.* **18**, 790–805.

Frydenberg, M. (1990b) The chain graph Markov property. *Scan. J. Statist.* **17**, 333–353.

Fuller, W.A. (1987) *Measurement error models.* New York: Wiley.

Geiger,D. and Pearl, J. (1993) Logical and algorithmic properties of conditional independence. *Ann. Statist.* **21**, 2001-2021.

Geng, Z. and Asano, C. (1993) Strong collapsibility of association measures in linear models. *J. Roy. Statist. Soc. B* **55**, 741–747.

Giesen, H., Böhmeke, W., Effler, M., Hummer, A., Jansen, R., Kötter, B. Krämer, H.-J., Rabenstein, E. and Werner, R.R. (1981) *Vom Schüler zum Studenten. Bildungslebensläufe im Längsschnitt.* Monografien zur Pädagogischen Psychologie, **7**. Munich: Reinhardt.

Glonek, G.F.V. and McCullagh, P. (1995) Multivariate logistic models.

J. Roy. Statist. Soc. B **57**, 533–546.

Goldstein, H. (1995) *Multilevel statistical models,* 2nd edn. London: Edward Arnold.

Goodman, L.A. (1970) The multivariate analysis of qualitative data: interaction among multiple classifications. *J. Amer. Statist. Assoc.* **65**, 226–256.

Goodman, L.A. (1971) Partitioning of chi-square, analysis of marginal contingency tables, and estimation of expected frequencies in multidimensional contingency tables. *J. Amer. Statist. Assoc.* **66**, 339–344.

Haavelmo, T. (1943) The statistical implications of a system of simultaneous equations. *Econometrica* **11**, 1–12.

Hardt, J. (1995) *Chronifizierung und Bewältigung bei Schmerzen.* Lengerich: Pabst.

Heath, A., Evans, G. and Payne, C. (1995) Modelling the class–party relationship in Britain, 1964–92. *J. Roy. Statist. Soc. A* **158**, 563–574.

Hogg, R.V. and Craig, A.T. (1978) *Introduction to mathematical statistics.* New York: Macmillan.

Holland, P.W. (1986) Statistics and causal inference (with discussion). *J. Amer. Statist. Assoc.* **81**, 945–970.

Isham, V. (1981) An introduction to spatial point processes and random fields. *Internat. Statist. Rev.* **49**, 21–43.

Jöreskog, K.G. (1981) Analysis of covariance structures. *Scan. J. Statist.* **8**, 65–92.

Jørgensen, B. (1993) *The theory of linear models.* London: Chapman & Hall.

Karlin, S. and Rinott, Y. (1980) Classes of orderings of measures and related correlation inequalities I. Multivariate totally positive distributions. *J. Mult. Anal.* **10**, 467–498.

Kauermann, G. (1996) On a dualization of graphical Gaussian models. *Scan. J. Statist.,* to appear.

King, G., Keohane, R.O. and Verba, S. (1994) *Designing social enquiry.* Princeton, NJ: Princeton University Press.

Kohlmann C.W., Krohne, H.W., Küstner E., Schrezenmeir, J., Walther, U. and Beyer, J. (1991) Der IPC-Diabetes-Fragebogen: ein Instrument zur Erfassung krankheitsspezifischer Kontrollüberzeugungen bei Typ-I-Diabetikern. *Diagnostica* **37**, 252–270.

Koster, J. (1996) Markov properties of non-recursive causal models. *Ann. Statist.,* to appear.

Kreiner, S. (1989) Analysis of multidimensional contingency tables by exact conditional tests; techniques and strategies. *Scan. J. Statist.* **14**, 97–112.

Laucht, M., Esser, G. and Schmidt, M.H. (1992) Verhaltensauffälligkeiten bei Säuglingen und Kleinkindern: ein Beitrag zu einer Psychopathologie der frühen Kindheit: ein Instrument zur Erfassung krankheitsspezifischer Kontrollüberzeugungen bei Typ-I-Diabetikern. *Diagnostica* **37**, 252-270.

Lauritzen, S.L. (1989) Mixed graphical association models (with discussion). *Scan. J. Statist.* **16**, 273-306.

Lauritzen, S.L. (1996) *Graphical models.* Oxford: Oxford University Press.

Lauritzen, S.L. and Wermuth, N. (1989) Graphical models for associations between variables, some of which are qualitative and some quantitative. *Ann. Statist.* **17**, 31-54.

Lauritzen, S.L., Dawid, A.P., Larsen, B. and Leimer, H.-G. (1990) Independence properties of directed Markov fields. *Networks* **20**, 491-505.

Lawley, D.N. and Maxwell, A.E. (1971) *Factor analysis as a statistical method,* 2nd edn. London: Butterworth.

Leber M. (1996) Die Effekte einer poststationären telefonischen Nachbetreuung auf das Befinden chronisch Schmerzkranker. Dissertation, Medical School, Universität Mainz.

Lehmann, E.L. (1990) Model specification. *Statist. Sci.* **5**, 160-168.

Leimer, H.-G. (1993) Optimal decomposition by clique separators. *Discrete Math.* **113**, 99-123.

Little, R.J.A. and Rubin, D.B. (1987) *Statistical analysis with missing data.* New York: Wiley.

Mardia, K.V., Kent, J.T. and Bibby, J.M. (1979) *Multivariate analysis.* London: Academic Press.

McCullagh, P. and Nelder, J.A. (1989) *Generalized linear models,* 2nd edn. London: Chapman & Hall.

Miller, A.J. (1990) *Subset selection in regression.* London: Chapman & Hall.

Patterson, H.D. and Thompson, R. (1971) Recovery of inter-block information when block sizes are unequal. *Biometrika* **58**, 545-554.

Pearl, J. (1988) *Probabilistic reasoning in intelligent systems.* San Mateo, CA: Morgan Kaufmann.

Pearl, J. & Wermuth, N. (1994) When can association graphs admit a causal interpretation? In P. Cheseman and W. Oldford (eds.) *Models and Data, Artificial Intelligence and Statistics IV.* New York: Springer.

Plummer, M. and Clayton, D. (1996) Estimation of population exposure in ecological studies (with discussion). *J. Roy. Statist. Soc. B* **58**, to appear.

Rosenbaum, P.R. (1995) *Observational studies.* New York: Springer.

Rothman, K.J. (1986) *Modern epidemiology.* Boston: Little & Brown.

Rubin, D.B. (1974) Estimating causal effect of treatments in randomized and nonrandomized studies. *J. Educational Psychology* **66**, 688-701.

Rubin, D.B. (1987) *Multiple imputation for nonresponse in surveys.* New York: Wiley.

Simpson, E. H. (1951) The interpretation of interactions in contingency tables. *J. Roy. Statist. Soc. B* **13**, 238–241.

Snedecor, G.W. and Cochran, W.G. (1967) *Statistical methods,* 6th edn. Ames: Iowa State University Press.

Sobel, M.E. (1995) Causal inference in the social and behavioral sciences. In G. Arminger, C.C. Clogg and M.E. Sobel (eds), *Handbook of statistical modeling for the social and behavioral sciences.* New York: Plenum.

Speed, T.P. and Kiiveri, H.T. (1986) Gaussian Markov distributions over finite graphs. *Ann. Statist.* **14**, 138–150.

Spiegelhalter, D.J., Dawid, A.P., Lauritzen, S.L. and Cowell, R.G. (1993) Bayesian analysis in expert systems (with discussion). *Statist. Sci.* **8**, 219–283.

Spiegelhalter, D.J. and Smith, A.F.M. (1982) Bayes factors for linear and log-linear models with vague prior information. *J. Roy. Statist. Soc. B* **44**, 337–387.

Spirtes, P. (1995) Directed cyclic graphical representations of feedback models. In P. Besnard and S. Hanks (eds), *Proceedings of the 11th conference on uncertainty in artificial intelligence.* San Mateo CA: Morgan Kaufmann.

Spirtes, P., Glymour, C. and Scheines, R. (1993) *Causation, prediction and search.* New York: Springer.

Stevens, S.S. (1950) *Handbook of experimental psychology,* Chapter 1. New York: Wiley.

Stoker, T.M. (1986) Aggregation, efficiency and cross-section regression. *Econometrica* **54**, 171–188.

Streit, R. (1996) Graphische Kettenmodelle mit binären Zielgrößen. Modellierung und Datenbeispiele in psychologischer Forschung. Lengerich: Pabst.

Thomas, D., Stram, D. and Dwyer, J. (1993) Exposure measurement error: influence on exposure–disease relationships and methods of correction. *Ann. Rev. Public Health* **14**, 69–93.

Weisberg, S. (1980) *Linear regression analysis.* New York: Wiley.

Wermuth, N. (1976) Analogies between multiplicative models in contingency tables and covariance selection. *Biometrics* **32**, 95–108.

Wermuth, N. (1980) Linear recursive equations, covariance selection and path analysis. *J. Amer. Statist. Assoc.* **75**, 963–972.

Wermuth, N. (1987) Parametric collapsibility and the lack of moderating effects in contingency tables with a dichotomous response variable. *J. Roy. Statist. Soc. B* **49**, 353–364.

Wermuth, N. (1992) On block-recursive regression equations (with discussion). *Braz. J. Prob. Statist. (Revista Brasileira de Probabilidade e Estatistica)* **6**, 1–56.

Wermuth, N. (1995) On the interpretation of chain graphs. *Proceed. 50th*

Sess. Internat. Statist. Inst., Beijing, Vol. 1, Invited papers, 415–429.

Wermuth, N. and Cox, D.R. (1992) On the relation between interactions obtained with alternative codings of discrete variables. *Methodika* **6**, 76–85.

Wermuth, N. and Cox, D.R. (1995) Derived variables calculated from similar joint responses: some characteristics and examples. *Comput. Statist. Data Anal.* **19**, 223–234.

Wermuth, N., Cox, D.R. and Pearl, J. (1994) Explanations for multivariate structures derived from univariate recursive regressions. Submitted.

Wermuth, N. and Lauritzen, S.L. (1983) Graphical and recursive models for contingency tables. *Biometrika* **70**, 537–552.

Wermuth, N. and Lauritzen, S.L. (1990) On substantive research hypotheses, conditional independence graphs and graphical chain models (with discussion). *J. Roy. Statist. Soc. B* **52**, 21–72.

Wermuth, N. and Rüssmann, H. (1993) Eigenanalysis of symmetrizable matrix products: a result with statistical applications. *Scan. J. Statist.* **20**, 361–367.

Whittaker, J. (1990) *Graphical models in applied multivariate statistics.* Chichester: Wiley.

Wold, H.O. (1954) Causality and econometrics. *Econometrica* **22**, 162–177.

Wright, S. (1921) Correlation and causation. *J. Agric. Res.* **20**, 162–177.

Yates, F. (1935) Complex experiments (with discussion). *Suppl. J. Roy. Statist. Soc.* **2**, 181–247.

Yates, F. (1951) The influence of *Statistical methods for research workers* on the development of the science of statistics. *J. Amer. Statist. Assoc.* **46**, 19–34.

Yule, G.U. (1900) On the association of attributes in statistics. *Phil. Trans. Roy. Soc. (London) A* **194**, 257–319.

Zellner, A. (1962) An efficient method of estimating seemingly unrelated regressions and tests for aggregation bias. *J. Amer. Statist. Assoc.* **57**, 348–368.

Zellner, A. and Huang, D.S. (1962) Further properties of efficient estimators for seemingly unrelated regression equations. *Internat. Econom. Rev.* **3**, 300–313.

Zhao, L.P. and Prentice, R.L. (1990) Correlated binary regression using a quadratic exponential model. *Biometrika* **77**, 642–648.

List of Figures

List of Tables

Author index

Subject index

Printed in the United States
by Baker & Taylor Publisher Services